Volume 4

CHEMICAL OXIDATION

TECHNOLOGIES FOR THE NINETIES

Edited by

W. Wesley Eckenfelder

Eckenfelder, Inc.

Alan R. Bowers

Vanderbilt University

John A. Roth

Vanderbilt University

PROCEEDINGS OF THE FOURTH INTERNATIONAL SYMPOSIUM
CHEMICAL OXIDATION: TECHNOLOGY FOR THE NINETIES
VANDERBILT UNIVERSITY
NASHVILLE, TENNESSEE
FEBRUARY 16–18, 1994

TECHNOMIC
PUBLISHING CO., INC.
LANCASTER · BASEL

Chemical Oxidation, Volume 4
a TECHNOMIC ®publication

Published in the Western Hemisphere by
Technomic Publishing Company, Inc.
851 New Holland Avenue, Box 3535
Lancaster, Pennsylvania 17604 U.S.A.

Distributed in the Rest of the World by
Technomic Publishing AG
Missionsstrasse 44
CH-4055 Basel, Switzerland

Printed in the United States of America
10 9 8 7 6 5 4 3 2 1

Main entry under title:
    Chemical Oxidation: Technologies for the Nineties, Volume 4

A Technomic Publishing Company book
Bibliography: p.

ISSN No. 1072-2459
ISBN No. 1-56676-489-0

HOW TO ORDER THIS BOOK
BY PHONE: 800-233-9936 or 717-291-5609, 8AM–5PM Eastern Time
BY FAX: 717-295-4538
BY MAIL: Order Department
Technomic Publishing Company, Inc.
851 New Holland Avenue, Box 3535
Lancaster, PA 17604, U.S.A.
BY CREDIT CARD: American Express, VISA, MasterCard
BY WWW SITE: http://www.techpub.com

TABLE OF CONTENTS

PREFACE

Chemical oxidation technologies are rapidly maturing into a wide variety of processes for the treatment of difficult waste streams, including wastewater, groundwater, hazardous waste and air. In addition, an industry has developed around the production and marketing of chemical oxidants, oxidation equipment, and entirely new treatment processes employing oxidation in new and unique configurations.

While oxidation processes have been under development for many years, much of the reported data are empirical. These data are often applicable to specific wastes and not broadly useful. Additionally, many new applications and processes are reported each year. This volume is the fourth proceedings of the international symposium on chemical oxidation processes applied to environmental problems. These papers represent the state of the art in theory, design, and practices of chemical oxidation processes.

W. W. Eckenfelder
A. R. Bowers
J. A. Roth

Influence of Chlorine-Free Pulp Bleaching on Wastewater Treatment Processes

As evident from the content of this conference, chemical oxidation is widely employed for the oxidation of specific organics in groundwaters and for the pretreatment of industrial wastewaters. In this application, the primary objective is detoxification and enhanced biodegradability of the organics present.

In recent years, chemical oxidation has also been employed in industry for purposes of pollution prevention and waste minimization. A notable example of this is the substitution of ozone, hydrogen peroxide and chlorine dioxide for chlorine in the bleaching sequence of pulp.

Chlorine bleaching of pulp has been practiced for over fifty years. In the past, this was not an environment problem since there was little effect on BOD removal which had been the primary focus of effluent compliance. In recent years, however, effluent COD, color and aquatic toxicity have become a significant part of most effluent permits. The effluent from a chlorine bleached effluent contains significant amounts of AOX (adsorbable organic halides) which are only partially biodegradable. These organic by-products of the chlorine bleaching sequence will contribute to the non-degradable COD in the effluent after biological treatment. The effluent COD comprises as much as 46 percent of the total non-degradable effluent from a conventional bleach plant following biological treatment as evidenced in Table 1.

Since COD is a permit parameter in many parts of the world, this high non-degradable COD can create considerable permit compliance problems. The performance summary for the removal of conventional pollutants and for AOX is shown in Table 2(1). As can be seen, while BOD and TSS in conventional biological treatment removal exceeds 90 percent, bot COD and AOX are only partially removed. AOX removal from eight mills, as shown in Table 3(2) illustrated the same problem (i.e., poor AOX removal). The effluent AOX after biological treatment has been shown to be directly proportional to the influent concentration as shown in Figure 1(2).

W. Wesley Eckenfelder, D.Sc., Eckenfelder, Inc., 227 French Landing Dr., Nashville, TN 37228

TABLE 1

EFFLUENT CHARACTERISTICS FROM A CONVENTIONAL BLEACHED KRAFT WASTEWATER

Influent		Effluent				
BOD (mg/L)	COD (mg/L)	BOD (mg/L)	COD (mg/L)	COD_{ND} (mg/L)	% non-deg	BOD_5/COD_D
308	1,153	7	575	535	46	0.5

TABLE 2

PERFORMANCE SUMMARY FOR CONVENTIONAL POLLUTANTS

Parameter	50th & (90th) Percentile Values		
	AS	FSB	ASB
COD Removal (%)	54 (65)	55 (78)	57 (68)
BOD Removal (%)	96 (98)	96 (98)	96 (98)
NH_4-N_{effl}(mg/L)	1.5 (10.1)	0.25 (5.8)	0.25 (4.5)
(NO_2+NO_3) - N_{effl}(mg/L)	1.4 (8.0)	4.3 (11.1)	8.0 (13.2)
VSS effl (mg/L)	32 (110	62 (200)	75 (200)

Observed Overall AOX Removal Performance

Parameter	560th Percentile & (90th/50th) Values		
	AS	FSB	ASB
Total AOX Removal (%)	22 (1.7)	43 (1.3)	40 (1.3)
Filt-AOX Removal (%)	28 (1.5)	48 (1.2)	45 (1.2)
(Non-filt-AOX/Total AOX)$_{effl}$ (%)	8 (1.9)	8 (1.9)	8 (1.9)
(Non-filt-AOX/TSS)$_{effl}$ (mg/g)	45 (2.8)	28 (1.9)	20 (2.6)

AS - activated sludge
FSB = facultative stabilization basin
ASB - aerated stabilization basin

TABLE 3

AOX REMOVAL THROUGH BIOLOGICAL TREATMENT

Mill Code	Process Type	Furnish	Primary Treatment	Activated Sludge SRT	ASB HRT	AOX Removal (%)
A	Kraft	sw/hw	yes	4 days	--	38
B	Sulfite	hw	no	8 days	--	53
C	Sulfite	sw/hw	yes	20 days*	1.5 days*	17
D	Kraft	32/hw	yes	--	7.0 days	30
E	Kraft	sw/hw	yes	--	7.0 days	38
F	Kraft	sw/hw	yes	--	13.0 days	42
G	Kraft	sw/hw	yes	--	7.0 days	27
H	Kraft	sw/hw	yes	--	7.0 days	34

Note: SRT = solids retention time; HRT = hydraulic retention time; sw = softwood; hw = hardwood
*Only bleach caustic filtrate is treated by activated sludge; all combined wastewaters are then treated by ASB.

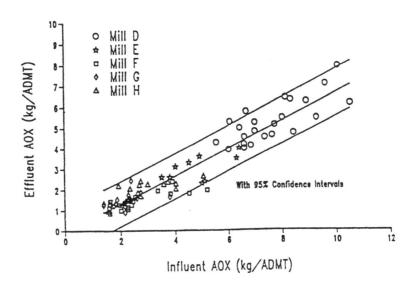

Figure 1. Correlation of Influent adn Effluent AOX Loadings Across ASBs. (Mill shutdown months excluded.) (2)

3

There is less AOX generated when pulps are delignified to lower the kappa number and more $C10_2$ is used in the chlorination stage. The use of 100 percent $C10_2$ in the chlorination stage reduced the AOX from 4.1 to 0.5 kg AOX/air dried ton pulp (3). Oxone can also be used to bleach pulp. A comparison of effluent characteristics using chlorine and ozone is shown in Table 4(4).

When hydrogen peroxide was used in the extraction stage, total available chlorine was reduced by 20 to 30 percent, color by 50 percent and AOX by 42 percent (5). In addition to reducing the amount of pollutant generated in the bleaching process, the biodegradability of the bleach plant effluent should also be considered. Activated sludge performance comparing conventional and oxygen bleaching is shown in Table 5(6). As can be seen, both the initial concentration of AOX is reduced and the percent AOX reduction increased through conventional biological treatment. In another case, conventional chlorine bleaching was replaced by hydrogen peroxide bleaching for a sulfite pulping mill. The reaction rate, K, with hydrogen peroxide bleaching was 52/day compared to 20/day at a comparable temperature with chlorine bleaching. Since this wastewater consists largely of acetate, furfural and methanol, it can be hypothesized that chlorination by-products exert some inhibition on the biooxidation process.

CONCLUSIONS

The substitution of chlorine dioxide, hydrogen peroxide or ozone for chlorine in pulp bleaching can significantly reduce the generation of non-degradable AOX. This AOX contributes to the effluent COD and in many cases to effluent aquatic toxicity. It has also been shown that the biodegradation rate of the organics in the wastewater may be enhanced.

REFERENCES

1. Hall, E.R. and Randle, W.G., Water Science Technology, 26, 1-2, 387-396, 1992.
2. Bryant, C.W., et. al. Water Science Technology, 26, 1-6, 417-425, 1992.
3. Liebergott, N. et al., "Lowering AOX Levels in the Bleach Plant", Proceedings 1992 TAPPI Environmental Conference.
4. Trimble, D.S., "Environmental Benefits of Ozone Based Bleaching", 1993 TAPPI Environmental Conference.
5. Klein, R.J., et al., "Hydrogen Peroxide Reinforced Extraction Lowers Chlorinated Organics and Color in Bleach Plant Effluent", 1991 TAPPI Environmental Conference.
6. Nevalainen, J., et al., Water Science Technology, 24, No. 3-4, pp. 427-430, 1991.

TABLE 4

COMPARISON OF CONVENTIONAL (CHLORINE) AND OZONE BLEACHING OF PULP

	CEDED	OZ(EO)D
	Kg/ADT	
BOD$_5$	16	2
COD	65	6
Color	185	1.5
Chloride	59	3.5
AOX	6.5	0.08
Water Use m^3/ADT	55.4	13.7

TABLE 5

COMPARISON OF CONVENTIONAL AND OXYGEN BLEACHING ON ACTIVATED SLUDGE PERFORMANCE

Average Concentrations of AOX (mg/L) and reduction percentages

	Influent	Effluent	% Reduction
Conventional	136	109	22
Oxygen	57	34	40

Average Reduction of Chlorinated Phenols (%)

	Phenols	Guajacols	Catacols
Conventional	39	41	50
Oxygen	45	79	63

SANTIAGO ESPLUGAS
DAVID F. OLLIS

Process Integration Development: Reactor Kinetic Models for Sequential Chemical and Biological Oxidation for Water Treatment

ABSTRACT

Of the thirty or so reports in the literature dealing with attempts to integrate a chemical followed by a biological oxidation process for water treatment, none contain serious attempts at constructing credible models of the pertinent chemical and biological kinetics in order to ascertain that (1) a reasonable predictive understanding of the (two) processes exists and (2) that such a combined kinetics model approach can be used to search conveniently for the optimal integrated system subject to appropriate operating and economic constraints.

This paper reports on such a model and its predictions for a (photo)chemical oxidation process for water remediation followed by a corresponding biological oxidation.

Calculated results provide conversion efficiencies versus fractional holding time in the (photo)chemical reactor at fixed total holding times, as well as the predicted performance contribution of each reactor to the overall conversion achieved. An optimum apportioning of reactor volumes (or, therefore, for a related problem, of capital and operating costs) is shown to exist. The breadth of such an optimum suggests that a considerable portion of the conversion task may be accomplished in the bioreactor.

Santiago Esplugas, Department of Chemical Engineering, University of Barcelona, Marti i Franques, 1 08028 Barcelona, Spain

David F. Ollis, Department of Chemical Engineering, North Carolina State University, Raleigh, NC 27695, USA

7

INTRODUCTION

The latest generation of chemical oxidation technologies, based in the use of ultraviolet (UV) radiation and/or chemical oxidants (UV-ozone, UV-hydrogen peroxide, UV-photocatalysis and ozone-peroxide) named "**Advanced Oxidation Processes**" (AOPs) may achieve complete mineralization of oxidizable contaminants of water (pesticides, solvents, fuels, halogenates, etc..). Each AOP produces hydroxyl radicals which are able to destroy even recalcitrant compounds such as chlorinated hydrocarbons. However, treatment costs are often greater than that for typical biological oxidation processes (e.g., activated sludges), where applicable. Previous studies have illustrated a higher degree of degradation when a chemical or photochemical treatment is combined with a biological treatment. In consequence, it is interesting to integrate these two processes in a reactor model in order to explore their potential combined efficiency for water treatment.

Kong and Sayler (1983) found that the simultaneous use of a sunlamp with natural river sediment, taken from PCB contaminated areas, enhanced the mineralization rate of 4CB (4-chlorobiphenyl) by 400% versus sediment alone, as Judged by C^{14} released as CO_2 from ring-labeled substrate over three day exposure [7].

Kearney et al. (1983) studied degradation of C^{14} labeled TNT (2,4,6-trinitrotoluene). They identified different products of TNT degradation and concluded that "soil metabolism, as measured by metabolic CO_2 evolution increased as the time of prior UV-O_3 treatment increased" [6].

Baxter and Sutherland (1984) examined the reverse sequence of microbial followed by photochemical treatment of 2, 4'-dichlorobiphenyl using first a pseudomonad isolated from activated sludge and then pyrex-filtered, mercury-xenon lamp light (spectral distribution similar to solar). The integrated sequence provided products which carried oxidation slightly further than the individual processes [3].

Katayana and Matsumura (1991) examined the simultaneous treatment by ultraviolet radiation (300 nm) and a UV resistant white rot fungus (*Phanerochaete chrysosporium* BKM F-1767) for the degradation of C^{14} labeled TCDD (2,3,7,8-tetrachlorodibenzo-p-dioxin), finding that simultaneous application of microbial and UV treatment provided more rapid release of C^{14} carbon dioxide than either UV or microbial treatment alone. They obtained similar positive results for individual dilute solutions of DDT, dieldrin, heptachlor, 3,4,3',4'-tetracholrobiphenyl and toxaphene. Several compounds were degraded to below the HPLC detection limits of 1-10 ng/mL in time periods of one to two weeks [5].

The objective of this paper is to provide a simple predictive system which can estimate the total conversion of individual reactants and a key intermediate as a function of holding times in a first, chemical or photochemical oxidation and a subsequent bioreactor. Our simple model contain a primary pollutant **A** which may be (photo)chemically degraded to an intermediate product **S**, but which is impossible to degrade biologically (very recalcitrant). The intermediate photoproduct may be degraded to CO_2 by continued chemical oxidation, and is also biodegradable and constitutes the primary carbon substrate for the bioreactor. According to this simple model, there will be two photochemical reactions

$$A \rightarrow S$$
$$S \rightarrow CO_2$$

and one biochemical reaction

$$S \rightarrow CO_2$$

It is assumed that there is no biodegradation of the main contaminant A.

PHOTOCHEMICAL TREATMENT

The chemical rate, r (g/(L.h)) for some homogeneous and most heterogeneous (photo)oxidations of organic pollutants in effluents normally follow equations of the LangmuirHinshelwood form, especially for photo-degradation (Braun, 1986) [4] (Turchi, 1990) [10] (Al-Ekabi, 1988) [1] (Matthews, 1987) [8] (Ollis, 1991) [9],

$$r = \frac{k \, K_{abs} C}{1 + K_{abs} C} \tag{1}$$

where C is the concentration of the pollutant, and model constants k and K_{abs} are the rate constant and binding constant, respectively. The rate constant k also includes the radiation absorption by the reaction medium. To simplify the mathematical model, a reaction first order with respect the concentration of the contaminant will be assumed (thus $r=-k_A C_A$). This form will characteristically be valid for dilute solutions of initial interest. For a continuous stirred tank reactor (CSTR) and first order kinetics with respect to the compound A, the concentration (C_{Ap}) Of A which leaves the (photo)reactor is as follows

$$C_{Ap} = \frac{C_{AO}}{1 + k_A \theta_p} \tag{2}$$

where θ_p -- liquid phase holding time = F(vol/time)/ V(volume).

Assuming also a first order (photo)degradation with respect to the intermediate S, the photoreactor outlet concentration, C_{sp}, is given by:

$$C_{Sp} = \frac{e k_A C_{Ao} \theta_p}{\left(1 + k_A \theta_p\right)\left(1 + k_s \theta_p\right)} \tag{3}$$

where C_{AO} is the inlet concentration of A (g/L), and k_A and k_s are the respective first order kinetic constants for the (photo)chemical conversions of A and S(h[-1]) respectively. The parameter e is a stoichiometric coefficient, generally less than one for this single intermediate example, which represents the grams of S formed per gram of A (photo)degraded. Because the later biological treatment conventionally uses mass variables (cell, substrate, etc.), it is convenient to proceed with mass concentration rather than molar units in the photoreactor as well.

The (photo)degradation yield (η_p) may be measured as the grams of original and intermediate contaminants eliminated in the (photo) reactor divided by the grams entering the (photo) reactor. That is:

$$\eta_P = \frac{C_{AO} - (C_A + C_s)}{C_{AO}} \qquad (4)$$

If the (photo)oxidation of intermediate compounds S does not occur, the (photo)degradation yield will be

$$\eta_P = \frac{C_{AO} K_A \theta_P (1 - e)}{1 + K_A \theta_P} \qquad (5)$$

and will equal zero when e = 1.

The only difference between the mathematical model of a chemical vs. photochemical oxidation reactor here is the dependence of kinetic constants k_A and k_s on radiation that enters the reactor (wave-length and flux) and that is absorbed by the medium. In general, k_A and k_s may both depend on concentrations C_A and C_s. Here, for simplicity, it is assumed that there is no relation between the two kinetic constants and no dependency on concentrations (true first order reactions) or spatial position.

BIOCHEMICAL TREATMENT

In the bioreactor, biological oxidative degradation of chemical compounds by the microorganisms takes place, usually resulting in cell growth and cell maintenance. A bioreactor description thus requires balances on both the cells (biomass) and the reactant (substrate). To simplify the model, it is assumed that the compound S formed in the photoreactor is the only substrate for the bioreactor and that the compound A leaving the photoreactor is not biodegraded nor does it affect the biochemical kinetics. More complex possibilities are considered in a subsequent paper.

A common functional relationship between the specific cell growth rate $\mu (h^{-1})$ and the concentration of the essential compound (substrate) is the Monod equation:

$$\mu = \frac{\mu_{MAX} C_S}{K_S + C_S} \qquad (6)$$

where μmax is the cell maximum specific growth rate C, and K_S (g/L) is the value of the substrate concentration at which the specific growth rate is half of its maximum value.

Denoting q (L/h) as the effluent flow rate and V_b (L) as the available bioreactor liquid phase volume, and assuming that the feed does not contain cell mass (sterile feed), the steady-state cell mass balance is production = removal, i.e.,

$$\frac{\mu_{MAX} C_{SO}}{K_S + C_{SO}} X V_b = qX \qquad (7)$$

where X (g/L) is the cell mass concentration and C_{Sb} is the substrate concentration in the well stirred bioreactor. The substrate degradation rate is related to this cell growth by the apparent yield factor Y defined as:

$$Y = \frac{\text{mass of cell formed}}{\text{mass of substrate consumed}} \qquad (8)$$

giving the following expression for the steady-state Substrate mass balance:

$$q\, C_{SP} - \frac{1}{y} \frac{\mu_{MAX}\, C_{Sb}}{K_S + C_{Sb}} X\, V_b = q\, C_{Sb} \qquad (9)$$

Equations 6 and 8 are solved for X and Csb to yield:

$$C_{Sb} = \frac{K_S}{\mu_{MAX}\, \theta_b - 1} \qquad (10)$$

$$X = Y\left(C_{Sp} - \frac{K_S}{\mu_{MAX}\, \theta_b - 1} \right) \qquad (11)$$

where θ_b is the liquid phase residence time in the bioreactor. The inverse of this last quantity $(1/\theta_b)$ is the dilution rate, D, equal to q/V, the number of tank liquid volumes which pass through the vessel per unit time.

The volumetric cell production rate (R) must equal the cell output flow rate (X/θ_b) in order to achieve steady state for the cell mass.

A biodegradation yield (η_b) may be defined as the amount of grams eliminated in the bioreactor divided by the grams entering the (photo)reactor-bioreactor system. That is,

$$\eta_b = \frac{C_{Sp} - C_{Sb}}{C_{AO}} \qquad (12)$$

and the global yield obtained, η_g, will be

$$\eta_g \frac{C_{AO} - C_{Sb}}{C_{AO}} = \eta_p + \eta_b \qquad (13)$$

LIMITING OPERATION CONDITIONS

For any culture following Monod kinetics, equation (10) indicates that the substrate (intermediate) concentration leaving the system in steady state operation will depend only on the residence time (θ_b) in the bioreactor. As θ_b decreases, the substrate concentration leaving the bioreactor increases. There is a limiting minimum value for θ_b (maximum D) given by a 'wash-out" condition corresponding to a zero cell mass in the bioreactor (Bailey and Ollis, 1986):

$$\theta_{b,limit} = \frac{K_S + C_{Sp}}{\mu_{MAX} C_{Sp}} \tag{14}$$

Because C_{Sp} depend on the residence time in the photoreactor (eq. 3), the limiting minimum value for θ_b will depend on θ_p, i.e., on photoreactor operation. Figure 1a shows this variation between $\theta_{b,limit}$, and θ_p for one example case ($\mu_{MAX} = 1$, $K_S = 0.2$, $K_A = 0.1$, $K_S = 0$, $e=1$, $C_{AO}=1$). The corresponding global time appears in Figure 1b.

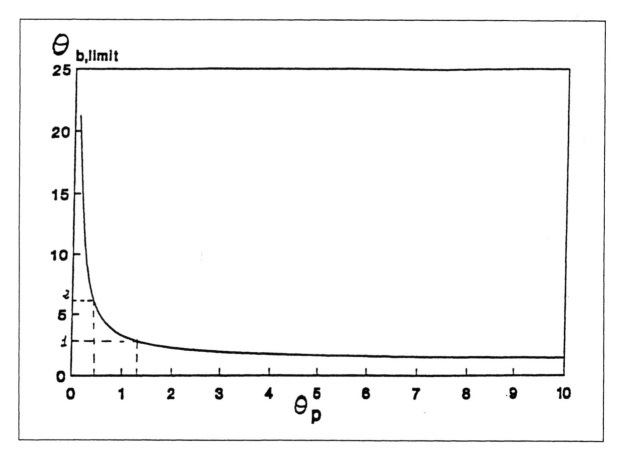

FIGURE 1a
UPPER LIMIT VALUE FOR THE BIOREACTOR RESIDENCE TIME

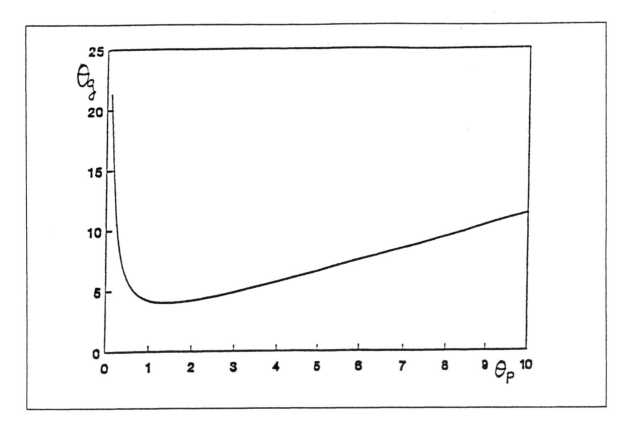

FIGURE 1b
CORRESPONDING GLOBAL RESIDENCE TIME $(K_A = 0.1; k_s = 0)$

With the added constraint of a fixed global residence time, θ_g

$$\theta_p + \theta_b + \theta_g = Constant \qquad (15)$$

there are now two limiting washout values of θ_b that give zero concentration of mass cell, X. The second washout corresponds to θ_b near zero; here as the photoreactor residence time approaches zero, insufficient substrates will be produced to yield cell mass in steady-state operation(washout at small C_s). When the photodegradation of the intermediate compound S is small, only a small global degradation will be reached; i.e., when θ_b nears the global residence time θ_g, only a small amount of substrate is produced photochemically (small η_p) and only a small global degradation will be achieved. In consequence there will be an intermediate, optimum residence time in the bioreactor, θ_b, where θ_b, min $< \theta_b < \theta_b$, lim) which gives a maximum in the combined or global degradation of contaminants (η_g).

MODEL ANALYSIS

Figure 2 summaries the simplified mathematical model for the (photo)chemical plus biological reactor system. There are nine degrees of freedom: initial concentration of the

13

pollutant (C_{AO}) stoichioimetric coefficients (e and Y), kinetic constants for (photo) reactions $(k_A$ and $k_s)$ and biological reaction $(K_S$ and $\mu_{MAX})$, and the liquid residence times in the photoreactor and bioreactor $(\theta_p$ and $\theta_b)$.

CONTAMINANT A:

$$C_{Ap} = \frac{C_{AO}}{1 + k_A\theta_p}$$

INTERMEDIATE S
(leaving photoreactor):

$$C_{Sp} = \frac{ek_A C_{AO}\theta_p}{(1 + k_A\theta_p)(1 + k_s\theta_p)}$$

INTERMEDIATE S
(leaving bioreactor):

$$C_{SO} = \frac{K_S}{\mu_{MAX}\theta_b - 1}$$

BIOMASS X

$$X = Y\left(C_{Sp} - = \frac{K_S}{\mu_{MAX}\theta_b - 1}\right)$$

EQUATIONS:..4

VARIABLES: Concentrations $(C_{AO}, C_{Ap}, C_{Sp}, C_{Sb}, X)$ 5
Stoichiometric coefficients (e, Y)..................................2
Kinetic constants $(k_A, K_S, k_S, \mu_{MAX})$ 4
Reactor operation (θ_p, θ_b)..................................2
TOTAL.....................13
DEGREES OF FREEDOM $(C_{AO}, e, Y, k_A, K_S, k_S, \mu_{MAX}, \theta_p, \theta_b)$.....................9

FIGURE 2
MODEL SUMMARY

RESULTS

A spreadsheet using the mathematical model described has been used to study the photo-bio-reactor system. Typical literature values for Monod parameters μ_{max} and K_s for activated sludge cultures range from 0.05 h^{-1} and 0.02 g/L to 1 h^{-1}and 0.4 g/L, respectively; for calculations, these values were fixed at $\mu_{max} = 1$ h^{-1} and $K_S = 0.2$ g/L. The concentration of the pollutant was set at $C_{AO} = 1$ g/L and the values for the stoichiometric coefficient e and the yield factor Y were set at 1. Typical rate constant values for first order photodegradation kinetics range from 0.01 h^{-1} to 100 h^{-1}.

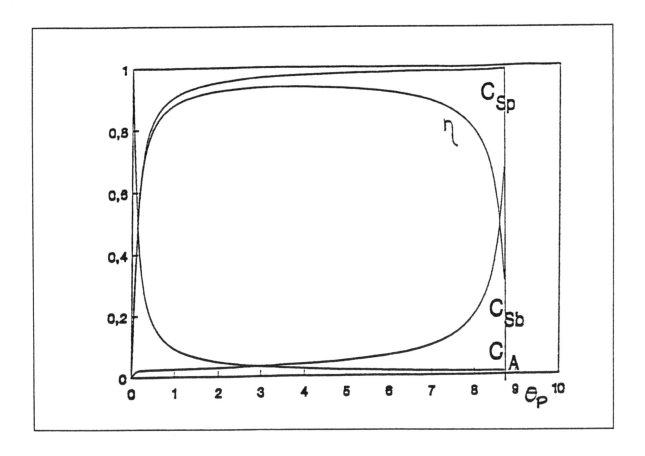

FIGURE 3
REACTANT AND INTERMEDIATE CONCENTRATIONS AND SYSTEM YIELD VS. RESIDENCE
TIME IN THE (PHOTO)REACTOR, θ_p (k_A =10, k_s=0, θ_g=10)

Figure 3 shows the calculated variation of the outlet feed contaminant concentration, C_{Ap}, the concentrations of substrate leaving the photoreactor, C_{sp}, and the bioreactor, C_{Sb}, and the global yield obtained, η_g, as a function of the residence time in the photoreactor. (Case: global residence time $\theta = 10$ hrs and photochemical kinetic constants $k_A = 10$ hr^{-1}, ks $= 0$ hr^{-1}

(no photoreaction of intermediate)). For the same conditions, Figure 4 shows the variation of the cell mass concentration in the bioreactor, X, and the cell mass growth rate, R. The results shown in Figures 3 and 4 establish that maximum η_g occurs at an optimum residence time in the photoreactor (θ_p) which does not correspond to a maximum cell growth rate. The global yield objective function is so broad that there is a wide range of θ_p values which give good results. In addition, because the stoichiometric factor e was fixed at 1 and the photochemical decomposition of S does not take place, the (photo)degradation yield, η_p is zero.

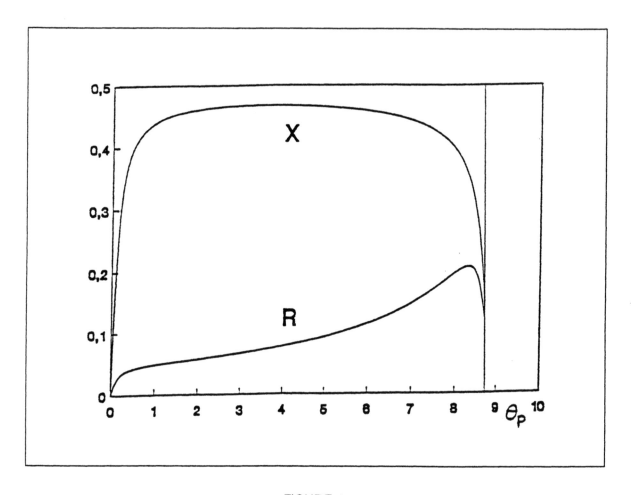

FIGURE 4
CELL MASS CONCENTRATION X AND CELL GROWTH RATE, R, VS. RESIDENCE TIME IN THE (PHOTO)REACTOR, θ_p (k_A =10, k_s =0, θ_g =10)

Figure 5 shows the influence of the photodegradation kinetic constant of the main pollutant (k_A) on the decomposition yield. As k_A increases, the maximum upper photodegradation yield also increases, and the limiting upper value of the residence time in the bioreactor, $\theta_{b,limit}$, increases slightly. For very small θ_p values the amount of substrate produced, C_{Sp} is too small to maintain cell in the steady-state, and wash-out occurs again.

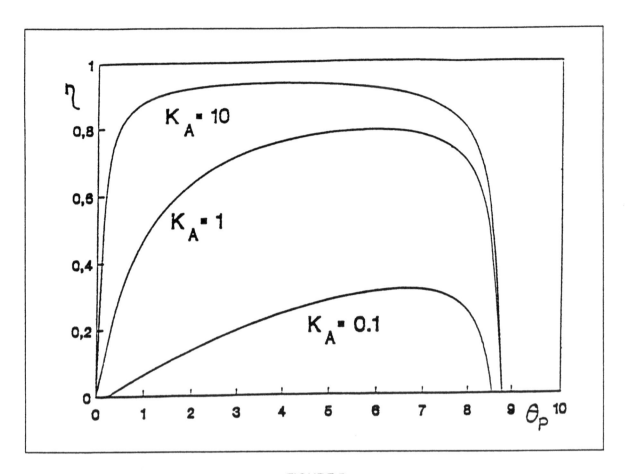

FIGURE 5
INFLUENCE OF THE PHOTODEGRADATION KINETIC CONSTANT OF THE PRIMARY
POLLUTANT ON THE GLOBAL DEGRADATION YIELD (k_s = 10, θ_g = 10)

Figures 6 and 7 show the influence of the (photo)degradation kinetic constant of the intermediate S substrate (case k_S = 1) on the different variables of the (photo)chemical-bioreactor system. The same curves are observed for η_g, and C_{Sb}; however a different curve appears for the substrate concentration leaving the photoreactor, C_{Sp}. In the previous case (k_S = 0), η_p= O and η_b = η_g; in the present case (k_S = 1), η_p >0 and the same value for η_g is obtained (Figure 8). It is interesting to observe the presence of a maximum of C_{Sp} typical for the reaction system A→S→*Products,* and an important reduction in the operation zone which can still maintain the bioreaction (the operation range for θ_p is now between 0.002 to 7.315).

17

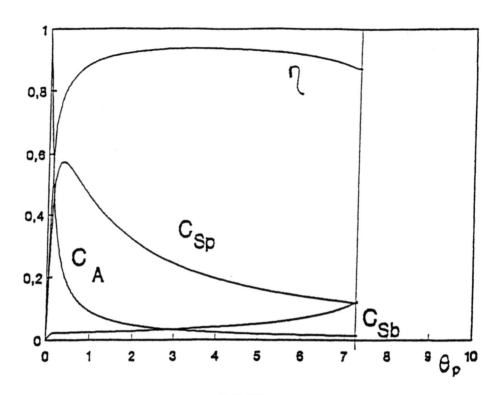

FIGURE 6
EFFECT OF THE (PHOTO)CHEMICAL RATE CONSTANT FOR SUBSTRATE (INTERMEDIATE)
ON CONCENTRATIONS AND THE SYSTEM YIELD, ($k_A = 10$, $k_s = 1.0$, $\theta_g = 10$)

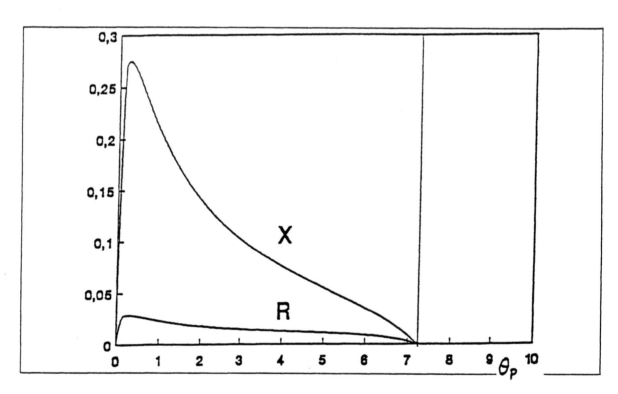

FIGURE 7
EFFECT OF THE (PHOTO)CHEMICAL RATE CONSTANT OF THE SUBSTRATE ON CELL MASS
CONCENTRATION X AND CELL

18

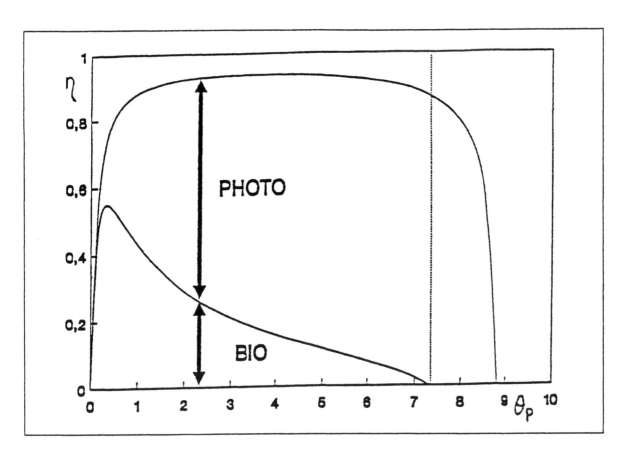

FIGURE 8
(PHOTO)CHEMICAL AND GLOBAL DEGRADATION YIELD WHEN THERE IS FINITE
(PHOTO)CHEMICAL DEGRADATION YIELD WHEN THERE IS FINITE (PHOTO)CHEMICAL
DEGRADATION OF THE (INTERMEDIATE) SUBSTRATE ($k_A=10$, $k_s=1.$, $\theta=10$)

As Table I shows, when the photochemical kinetic constant k_S increases, the range of the operation zone decreases.

TABLE I

K_s	$(\theta_{h\ limit})_1$	$(\theta_{h\ limit})_2$
0	1.202	9.998
1	2.685	9.998
2.5	4.169	9.998
5	5.653	9.998
10	7.136	9.998
25	8.620	9.998
50	9.293	9.998
100	9.680	9.998

CONCLUSIONS

There are two limiting or "wash-out" values for the residence time in the bioreactor, θ_b which depend on the residence time in the photoreactor θ_p.

As the kinetic constant for photodegradation of the intermediate compound increases, the range of values of possible residence times in the bioreactor , θ_b, decreases.

For a fixed global residence time, θ_g, there is an optimum residence time in the (photo)reactor, θ_p, which gives a maximum in the global yield of degradation, η_g. This optimum value does not correspond to the maximum cell growth rate.

For relatively fast (photo)chemical reactions, the optimum of the global degradation yield versus residence time in the photoreactor, θ_p is very broad. This broadness allows operation over a wide range of bioreactor vs. photoreactor residence times; in particular, it allows for maximal utilization of inexpensive bioreactor and minimal utilization of photoreactor residence time (thus volume, and thus cost).

NOMENCLATURE

C = concentration, $g.dm^{-3}$
e = stoichiometric coefficient. dimensionless
K = photochemical kinetic constant, h^{-1}
k = biochemical kinetic constant, $g.dm^{-3}$
q = volumetric flow, $dm^3.h^{-1}$
R = cell production rate, $g.dm^{-3}.h^{-1}$
r = photochemical reaction rate, $g.dm^{-3}.h^{-1}$
V = volume, dm^3
W = radiation flow, $einstein.s^{-1}$
X = cell mass concentration, $g.dm^{-3}$

GREEK LETTERS

θ = residence or spatial time, h^{-1}
μ = specific cell grow rate, h^{-1}
η = yield, dimensionless

SUBSCRIPTS

0 = initial
1 = minimum limit
2 = maximum limit
A = main pollutant
abs = absorbed
b = bioreactor
g= global

limit = limit condition
max = maximum
p = photoreactor
s = substrate (reaction intermediate)

REFERENCES

1. Al-Ekabi, H. and Serpone, N., "Kinetic studies in heterogeneous photocatalysis.", Langmuir 5, 250-255, (1989).

2. Bailey, J.E. and Ollis, D.F., "Biochemical engineering fundamentals", Mc Graw-Hill, New York (1986)

3. Baxter, R.M. and Sutherland, D.A., "Biochemical and photochemical processes in the degradation of chlorinated biphenyls" Environ. Sci. Technol., 18, 608-610 (1984)

4. Braun, A., Maurette, M.T. and Oliveros, E.,"Technologie photochimique", Presses Polytechniques Romandes, Laussane, (1986)

5. Katayana, A. and Matsumura, F., "Photochemically enhanced microbial degradation of environmental pollutants", Environ. Sci. Technol.,

6. Kearney, P.C., Zeng, Q. and Ruth, J.M., "Oxidative pretreatment accelerates TNT metabolism in soils", Chemosphere, 12, 1583-1597 (1983)

7. Kong, H-L. and Sayler, G.S., Degradation and total mineralization of monohalogenated biphenyls in natural sediment and mixed bacterial culture", Applied and Environmental Microbiology, 46, 666-672 (1983)

8. Matthews, R.W., "Photooxidation of organic impurities in water using thin films of titanium dioxide", Journal of Physical Chemistry, 91, 3328-3333 (1987) ....,

9. Ollis, D.F., Pelizzetti, E., and Serpone, N.,"Destruction of water contaminants", Environ. Sci. Technol., 25, 1523-1529 (1991)

10. Turchi, C.S. and Ollis, D.F., "Photocatalytic degradation of organic water contaminants: Mechanisms involving hydroxyl radical attack", Journal of Catalysis, 122, 178-192 (1990)

G. GORDON, G. PACEY
B. BUBNIS
S. LASZEWSKI
J. GAINES

Safety in the Workplace: Ambient Chlorine Dioxide Measurements in the Presence of Chlorine

ABSTRACT

A study was undertaken to measure and model ambient chlorine dioxide (ClO_2) concentrations in the presence of chlorine (Cl_2). The permissible exposure level (PEL) for ClO_2 as established by the Occupational Health & Safety Administration (OSHA), is 0.1 ppm (≈ 0.3 mg/m^3) and the short term exposure limit (STEL) is 0.3 ppm (≈ 0.9 mg/m^3). Measurements were made using ion mobility spectrometry (IMS) to quantitate ambient ClO_2 concentrations under laboratory "spill" conditions. Rates of emission or volatilization "downwind" at specified heights and distances from the spill were calculated.

Chlorine dioxide at equilibrium and 25 °C, is about 23 times more concentrated in aqueous solution or about 10 times more soluble in water than Cl_2 which it resembles in appearance and odor. The commercial generation of ClO_2 usually involves Cl_2, a fact that causes great confusion and inaccuracies in ambient ClO_2 measurements. The most readily available and commonly used "monitors" are unable to speciate ClO_2 in the presence of Cl_2.

IMS is a time of flight technique. Sample (ambient air or a process stream) is drawn through an inlet and permeates across a membrane (which provides a limited degree of selectivity) to a reaction cell. In the cell, the sample is ionized by a series of ion-molecule reactions under the influence of an electric field. The ions enter a "drift" region and are separated according to size and shape. Components are identified based on a drift time similar to gas chromatography but on the millisecond time scale.

Data will be presented describing the experiments, the model and its usefulness in predicting exposure levels to ambient ClO_2.

INTRODUCTION

Chlorine dioxide (ClO_2) is a selective and versatile oxidant which has been applied at many industrial worksites in the agriculture, food processing, iron and steel, pulp and paper, textiles, water purification and wastewater treatment fields. Chlorine dioxide has an Occupational Health and Safety Administration (OSHA) permissible exposure limit (PEL) of 0.1 ppm (≈ 0.3 mg/m^3) and a short term exposure limit (STEL) of 0.3 ppm (≈ 0.9 mg/m^3).

Gilbert Gordon, Gilbert Pacey, Bernard Bubnis, Steve Laszewski and Jeffrey Gaines, Department of Chemistry, Miami University, Oxford, OH 45056, USA

Acute exposure results in irritation to the eyes, nose, throat and lungs. Prolonged exposure can cause chronic bronchitis, reactive airways disease and pulmonary edema.

A study was undertaken to measure and model ambient ClO_2 concentration in the presence of Cl_2 under conditions similar to an event that occurred at a potato processing plant where chlorine dioxide was used in french fry mixing tanks. Over a four year period, employee exposure complaints were documented at the uncovered mixing tanks believed to contain 0.5 mg/L aqueous chlorine dioxide. Following a large spill incident, the chlorine dioxide system was removed and a lawsuit was brought by employees who had physician documented evidence of pulmonary injury. A series of small scale laboratory experiments were subsequently undertaken to evaluate the likelihood of employee injury from sufficient chlorine dioxide volatilization at the former mixing tanks.

At equilibrium and 25 °C, ClO_2 is about 23 times more concentrated in aqueous solution than in the gas phase or about 10 times more soluble in water than chlorine (Cl_2) which it resembles in appearance and odor. The generation of ClO_2 usually involves Cl_2, a fact that causes great confusion and inaccuracies in gas phase ClO_2 measurements since most monitors are unable to speciate ClO_2 in the presence of Cl_2. The laboratory measurement of ClO_2 in the presence of Cl_2 was made using ion mobility spectroscopy (IMS). Rates of emission or volatilization "downwind" at specified heights and distances from the "spill" were measured and volatilization half-lives were calculated using the two-film concept. An estimated volatilization half-life of 4.3 hours at the 2,000 liter mixing tanks was calculated corresponding to a chlorine dioxide emission rate of 2.5 x 10^{-5} g/sec (60 sec = 1.5 mg).

VOLATILIZATION MODELLING

Volatilization is affected by many factors[1] including the Henry's Law constant and the physical properties of aqueous solubility, vapor pressure, diffusivity coefficient and rate-controlling factors such as water body depth, presence of turbulence, sediment content and wind speed and stability. The calculation of rates of emission and volatilization half-life require the application of Henry's Law which states[2] that "the mass of a slightly soluble gas that dissolves in a definite mass of a liquid at a given temperature is very nearly directly proportional to the partial pressure of that gas".

The general procedure for calculating rates of volatilization[1] are as follows:

1. Calculate the Henry's Law Constant (H). If H>3×10^{-7}, the chemical can be considered volatile.

2. Determine the nondimensional Henry's Law Constant (H').

    $$H' = H/RT$$

3. Compute the liquid-phase exchange coefficient (k_1).

    $$k_1 = 23.51(V_c^{0.969}/Z_{0.673})(32/M)^{\frac{1}{2}} \text{ cm/hr}$$

where: V_c = liquid flow velocity
Z = liquid depth
M = molecular weight

4.   Compute the gas-phase exchange coefficient (k_g).

$$k_g = 1137.5(V_w + V_c)(18/M)^{\frac{1}{2}} \text{ cm/hr}$$

where: Vw = wind velocity above liquid
V_c = liquid flow velocity
M = molecular weight

5.   Compute the overall liquid-phase mass transfer coefficient (k_L).

$$k_L = \frac{(H/RT)k_g k_1}{(H/RT)k_g + k_1} \quad \text{cm/hr}$$

6.   Compute the volatilization half-life ($\tau_{\frac{1}{2}}$)

$$\tau_{\frac{1}{2}} = 0.69 \, Z/k_L = 0.69/k_V$$

where $k_V = k_L/Z$ = volatilization rate constant

CHLORINE DIOXIDE DETECTION

Chlorine dioxide in the gas phase was measured using ion mobility spectroscopy. Figure 1 is a schematic of an IMS sample chamber[3]. Sample (ambient air or a process stream gas) is drawn into an inlet and the compounds of interest permeate across a membrane which provides a limited degree of selectivity.

Figure 1.   IMS Cell Schematic

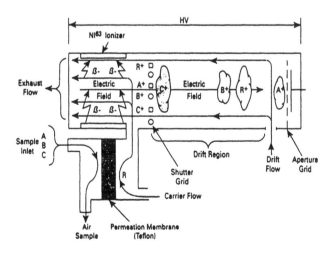

25

On the receiving side of the membrane, compounds are swept up by a carrier gas that delivers the sample to a reaction cell. In the cell, the sample is ionized by a series of ion-molecule reactions initiated by β particles emanating from a Ni^{63} source. These ions are attracted under the influence of an electric field to a shutter. The shutter opens and closes periodically to allow the ions into a "drift region" where they separate according to the size and shape of the specific ion and arrive at a collector grid at a unique "drift time". Components are identified based on a drift time (Figure 2) similar to gas chromatography but on the millisecond time scale[4]. Fixed point IMS is achieved by monitoring the unique drift time of the component of interest.

Figure 2. IMS "Drift" Spectrum of Chlorine Dioxide[5]

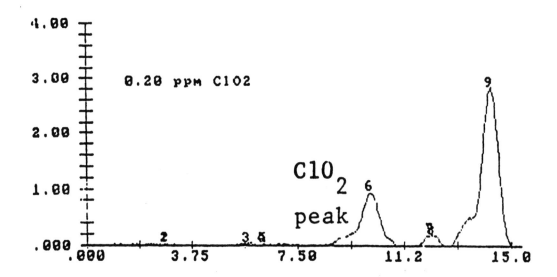

All IMS measurements were made using a Sensidyne/ETG fixed point ion mobility spectrometer calibrated with a Sensidyne ToxiCal chlorine dioxide gas generator. Real-time strip chart recordings were gathered using a Servogor 111 recorder interfaced to the spectrometer through a 4-20 mA current loop.

Chlorine dioxide solutions of concentrations up to 5000 mg/L were prepared[6,7] by mixing potassium persulfate with sodium chlorite in a two tower bubble chamber using nitrogen as a carrier gas. The solution concentrations were verified by UV measurement at 360 nm using short pathlength quartz cells and a Milton-Roy 3300 Diode Array Spectrophotometer.

Spill experiments were carried out in either a small pan (10 liter) or a 50 gallon stainless steel tank placed in an empty room measuring approximately 40' × 60' × 10'. Air flow above the tank was provided by a two-step circulating fan. Turbulent surface mixing was achieved using a variable speed paint mixer. Measurements taken at various heights above the spill tank were made by adjusting the IMS instrument to the desired height from a ceiling pulley.

LABORATORY "SPILL" EXPERIMENTS

A series of experiments was undertaken to measure the ambient chlorine dioxide concentration as a function of solution chlorine dioxide concentration, distance and height from the mixing tank, and surface turbulence. These data were used to calculate a measured chlorine dioxide volatilization half-life and rate constant which were then compared to predicted values.

Gas phase measurements described in terms of "parts per million" can be confusing. Generally the gas phase concentration is a volume-volume (v/v) relationship meaning that a 5 ppm (v/v) chlorine dioxide concentration is equivalent to 5×10^{-6} liters of chlorine dioxide in 1 liter of air or 0.276 mg of chlorine dioxide in m^3 of air.

Figure 3 shows ambient chlorine dioxide measurements taken 1 meter above the small 10 liter holding pan containing dilute chlorine dioxide solutions (<2 mg/L). Under highly concentrated conditions (200 - 860 mg/L), a similar linear effect was exhibited in the large tank experiments with measurement at 2 meters above the tank.

Figure 3.   Chlorine Dioxide Diffusion Into The Gas Phase At Low Concentration

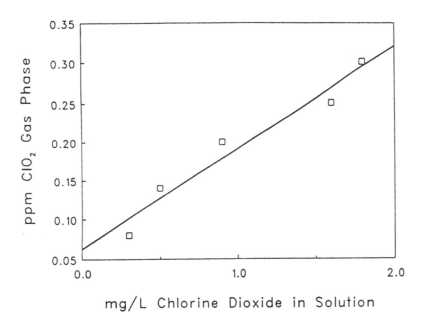

mg/L Chlorine Dioxide in Solution

Experiments were undertaken to determine the effect of distance from the mixing tank in the presence and absence of air motion above the tank. At increasing distances from the mixing tank, ambient measurements decreased. This was not unexpected considering the reactivity of chlorine dioxide and the potential for gas phase dilution to occur when measuring at points further away from the mixing tank. Upon the application of moving air above the mixing tank, ambient chlorine dioxide measurements also increased. Under "splash" conditions with moving air, two fronts of chlorine dioxide vapor were measured. The first vapor wave appeared to be higher in concentration and is postulated to be the product of a large initial release of chlorine dioxide at the time of the spill. The second more controlled vapor wave appeared to be under pseudo equilibrium conditions showing a gradual decrease

27

in ambient chlorine dioxide as the surface area of the tank settled. This same effect was also noticed in experiments where surface turbulence was added. Figure 5 shows a release profile of chlorine dioxide in an open tank with surface mixing.

Figure 4. Chlorine Dioxide Diffusion Into The Gas Phase At High Concentration

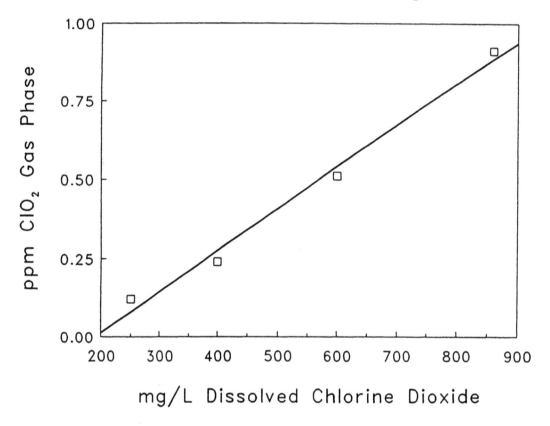

Figure 5. Impact of Surface Mixing on Chlorine Dioxide Release

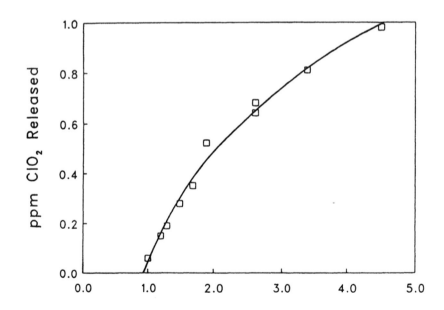

CALCULATED VOLATILIZATION PARAMETERS

The following parameters were calculated for chlorine dioxide.

Henry's Law Constant, H	4.0×10^{-2} atm m^3/mole
Nondimensional Henry's Law Constant, H'	1.7
Liquid-phase exchange coefficient, k_l	16.1 cm/hr
Gas-phase exchange coefficient, k_g	1549.2 cm/hr
Overall liquid-phase mass transfer coefficient, k_L	16.0 cm/hr
Rate constant, k_V	0.01 min^{-1}
Volatilization half-life, $\tau_{1/2}$	66 min

The chlorine dioxide volatilization rate constant is calculated as the slope of chlorine dioxide concentration *vs.* time as shown in Figure 6. The measured slope (0.024 ± 0.002)) *vs.* predicted (0.01) differed by only a factor of two at the 95% confidence level (predictive techniques are generally in agreement with actual processes within a factor of 10 and probably a factor of 2 or 3 in most cases). The corresponding measured half-life ($\tau_{1/2}$) was calculated to be between 25 -35 minutes.

Adjusting the calculations for the larger 2,000 liter mixing tanks at the potato plant, a $\tau_{1/2}$ of 4.3 hours was estimated giving a chlorine dioxide emission rate of 2.5×10^{-5} g/sec. This rate of emission coupled with dispersion modelling demonstrated the likelihood that ambient chlorine dioxide concentrations above the OSHA PEL and subsequent employee exposure at plant locations within 20 feet of the mixing tanks as shown in Figure 7.

Figure 6. Chlorine Dioxide Rate Constant Curves

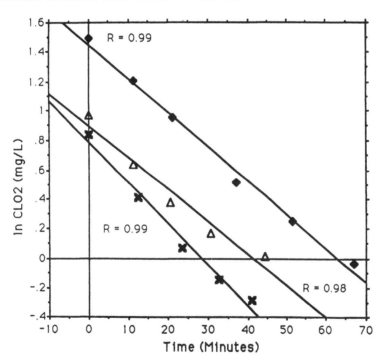

Figure 7.   Ambient Chlorine Dioxide Predictive Model

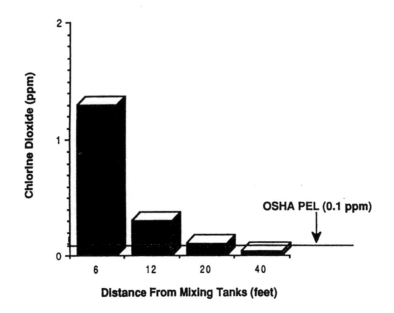

Literature

1. Thomas, R.G. Handbook of Chemical Property Estimation Methods, Lyman, W.; Reehl, W.; Rosenblatt, D. (Editors), **1982**, McGraw-Hill.

2. Handbook of Chemistry and Physics, 56th Edition, **1975**, CRC Press, p. F-103.

3. Bacon, A.T.; Getz, R.; Reategui, J. "Ion-Mobility Spectrometry Tackles Tough Process Monitoring", Chemical Engineering Progress, **1991**, June.

4. Spangler, G.E.; Carr, T.W. Instrument Design and Description in "Plasma Chromatography", **1984**, Plenum, New York.

5. Bacon, T. "Real Time Monitoring of Chlorine Dioxide Without Interference from Chlorine", Pittsburgh Conference Paper 326P, **1992**, New Orleans, LA.

6. Granstrom, M.L.; Lee, G.F. "Generation and Use of Chlorine Dioxide in Water Treatment", J. Am. Water Works Assoc., **1958**, 50, 1453-1466.

7. Wajon, J.E.; Rosenblatt; D.H.; Burrows, E.P. "Oxidation of Phenol and Hydro-quinone by Chlorine Dioxide", Environ. Sci. Technol., **1982**, 16, 396-402.

HARRY M. CASTRANTAS
JAMES L. MANGANARO

Hydrogen Peroxide and Caro's Acid — Powerful Oxidants for Cyanide Detoxification: A Review and Case Study

ABSTRACT

A brief review of the use of peroxygens to destroy free CN^-, WAD (weakly acid dissociable) CN^- and tightly complexed cyanides will be followed by a case history on the use of Caro's acid (peroxymonosulfuric acid) to treat cyanide in a tailings slurry at a large North American Gold mine.

Peroxygens such as hydrogen peroxide and peroxymonosulfuric acid (Caro's acid) will destroy cyanides. The choice of peroxygen depends on the type of cyanide and the nature of the substrate. Free cyanide and some WAD cyanides are destroyed with catalyzed hydrogen peroxide. For highly complexed cyanides, an advanced oxidation process such as UV-H_2O_2 is required. For treating free and WAD CN^- in slurries containing high concentrations of heavy metals such as gold mine tailings, Caro's acid is most effective.

Caro's acid (peroxymonosulfuric acid) was used to successfully treat a 6,500 GPM gold mining tailings slurry. Using an on-site FMC generator, Caro's acid reduced the CN^- concentration from 40 ppm to below the target level of 20 ppm. At increased Caro's acid levels, CN^- levels as low as 4 ppm were achieved. The raw material costs for the detoxification were reduced over 60% with Caro's acid compared to using H_2O_2 alone. In addition to the considerable savings in chemical raw material costs, the detoxification time with Caro's acid was much less compared to H_2O_2.

PEROXYGENS AND CYANIDE DESTRUCTION

Free CN^-, weakly complexed cyanides and highly complexed cyanides can be destroyed with hydrogen peroxide or Caro's acid. The peroxygen and specific conditions to be used depends on the type of cyanide and substrate to be treated.

For solutions containing free and weakly acid dissociable CN^-, H_2O_2 catalyzed with a soluble copper salt is the preferred treatment. A vanadium, tungsten or silver salt may be substituted for the copper salt.

Harry M. Castrantas and James L. Manganaro, FMC Corporation, Peroxygen Chemical Division, Box 8, Princeton, NJ 08543, USA

$$\text{H}_2\text{O}_2 + \text{CN}^- \xrightarrow[\text{pH 9-11}]{\text{Cu}^{++}} \text{CNO}^- + \text{H}_2\text{O}$$

Where the use of copper as a catalyst may be objectionable, citric acid can be substituted [1].

For tightly bound complexes such as ferrocyanides, the Fe-CN bond must first be broken before the CN^- can be attacked. This can be accomplished by using UV to dissociate the Fe-CN^- bond and H_2O_2 to destroy the free CN^- [2].

$$\text{Fe(CN)}_6^{-4} \xrightarrow{\text{UV}} \text{Fe}^{++} + 6\ \text{CN}^-$$

$$\text{CN}^- + \text{H}_2\text{O}_2 \longrightarrow \text{CNO}^- + \text{H}_2\text{O}$$

The Bureau of Mines developed a process for removing cyanides and heavy metals. H_2O_2 and sodium thiosulfate are used to convert free CN^- and WAD CN^- to thiocyanate (SCN^-). Ferrocyanide is removed by adding steryldimethylbenzyl ammonium chloride. Other heavy metals are removed by precipitation with ferric sulfate [3].

Slurries such as tailings from gold mining operations usually contain heavy metals which can lead to excessive decomposition of H_2O_2 when used to treat free CN^- and WAD CN^-. This reduces the amount of the H_2O_2 for cyanide destruction. In these cases, the use of Caro's acid is preferred since (1) Caro's acid is less sensitive to decomposition from heavy metals than H_2O_2. (2) Caro's acid reacts substantially faster than H_2O_2 with cyanides thereby minimizing the metal induced peroxygen decomposition reaction.

$$\text{H}_2\text{SO}_5 + \text{CN}^- \xrightarrow{\text{Fast}} \text{H}_2\text{SO}_4 + \text{CNO}^-$$

Ammonium, potassium or sodium persulfate will also destroy cyanides but react much slower compared to Caro's acid.

$$\text{Na}_2\text{S}_2\text{O}_8 + \text{CN}^- + \text{H}_2\text{O} \longrightarrow \text{CNO}^- + 2\ \text{NaHSO}_4$$

The cyanate (CNO^-) formed from the reaction of cyanides with peroxygens hydrolyzes as follows:

$$\text{CNO}^- + 2\ \text{H}_2\text{O} \longrightarrow \text{NH}_4^+ + \text{CO}_2$$

In addition to its use in gold mines, the focus of this paper, Caro's acid is applicable to destroying cyanides in a range of waste discharges, including electroplating operations, iron and steel operations, organic chemicals, and plastics.

PREPARATION AND PROPERTIES OF CARO'S ACID

Caro's acid is formed as follows:

$$H_2SO_4 + H_2O_2 ----------> H_2SO_5 + H_2O$$

Caro's acid can be prepared from concentrated hydrogen peroxide and concentrated sulfuric acid. Caro's acid yield will drop if dilute H_2O_2 and acid are used in its preparation. The exothermic nature of the reaction is primarily due to the heat of dilution of the sulfuric acid. The temperature rise of 50°C-60°C that results, intensifies the inherent instability of Caro's acid. For this reason, it is generated and consumed immediately at the point of use.

For laboratory use, small quantities of Caro's acid may be prepared by slowly adding H_2O_2 to concentrated sulfuric acid contained in a inert container e.g., glass, surrounded by an ice bath. Prepared in this way, the Caro's acid can be stored under refrigeration for several days.

Caro's acid also may be prepared by adding sulfuric acid to aqueous slurries of sodium, ammonium or potassium persulfate [4].

$$Na_2S_2O_8 + H_2O \xrightarrow{\quad H_2SO_4 \quad} NaHSO_4 + NaHSO_5$$

Caro's acid solutions are clear, colorless and have an oily consistency. Caro's acid possesses the properties of an oxidizer and strong acid. A typical solution prepared from 70% H_2O_2 and 93% sulfuric acid at a 2.5/1 mole ratio of H_2SO_4 /H_2O_2 will contain 25% Caro's acid, 47 % sulfuric acid, 3% hydrogen peroxide and 25% water. Caro's acid once formed and diluted only slowly hydroyzes back to sulfuric acid and H_2O_2.

In addition to detoxifying cyanides, Caro's acid can destroy phenols, alcohols, aldehydes and ketones.

GOLD MINING CASE STUDY

OVERVIEW OF GOLD MINE MILLING OPERATION

In a typical gold mining milling operation, ore is dry crushed followed by wet grinding. The aqueous slurry is sent to a series of leach tanks where sodium cyanide, a lixivant is added along with lime and air. After the gold solution is drawn off in a counter-current decantation process, the tailings containing residual cyanide is sent to a tailings pond for disposal (see Figure 1).

If the tailings require treatment to reduce residual cyanide concentrations, the detoxification agent is usually added to the tailings prior to discharge to the tailings pond. Alternatively the detoxification agent can be added directly to the pond.

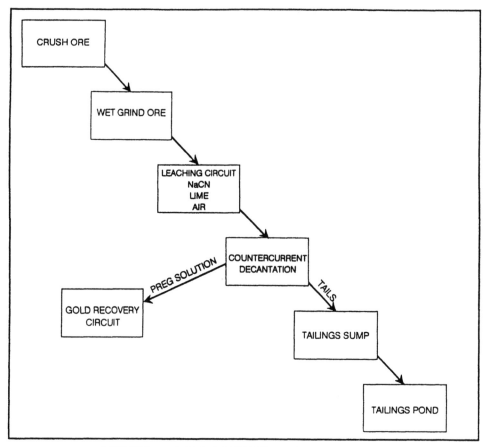

FIGURE 1 - GOLD MINE MILLING OPERATION

CYANIDE DETOXIFICATION BACKGROUND

A large North American gold mine processing 6,500 GPM tailings containing WAD CN^- residuals of 40-50 ppm had been using ferrous sulfate to complex the CN^- and reduce the hazard to wildlife in their tailings pond. These techniques did not prove entirely effective however and the mine staff switched from ferrous sulfate to H_2O_2 in the Spring of 1992. Hydrogen peroxide reduced the cyanide levels to 25 ppm, the target level.

As part of the mine's continuous improvement program, the staff's next goal was to reduce detoxification costs and improve the detoxification kinetics. With H_2O_2, a portion of the CN^- detoxification is completed within the 20 minute travel time to the pond from the tailings sump through HDPE lines to the tailings pond. The remainder of the detoxification is completed in the pond after several more minutes. Ideally, the entire detoxification should be completed before the treated tailings enters the pond. In this way, wildfowl landing on the pond in the vicinity of the tailings discharge lines would not be subject to high cyanide levels.

It was decided to evaluate Caro's acid (peroxymonosulfuric acid), a cyanide detoxification agent known to possess fast cyanide detoxification kinetics and a high detoxification efficiency.

34

CN⁻ DETOXIFICATION PROCESS DESCRIPTION

For the trial, an FMC Caro's acid generator was supplied which consisted of two components: (1) a feed skid containing the chemical feed pumps, a programmable logic controller and a water deluge system, (2) a reactor which was located over the tailings sump remote to the feed skid (see Figure 2).

Tank trucks of hydrogen peroxide and sulfuric acid served as temporary storage vessels during the trial.

The process described is continuous. Sulfuric acid and hydrogen peroxide are pumped from the feed skid at a tightly controlled rate and ratio of H_2SO_4/H_2O_2 to the CA reactor. The Caro's acid is then metered into the tailings stream where rapid oxidation of cyanide takes place. After the Caro's acid is mixed with the tailings slurry, it travels about one mile (20 minutes) prior to release to the tailings pond.

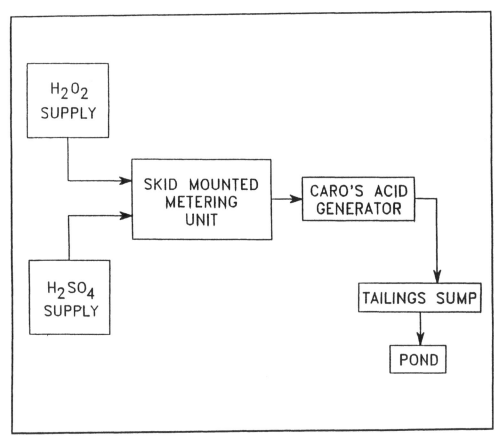

FIGURE 2 - SCHEMATIC FOR CARO'S ACID DETOXIFICATION
OF CYANIDE IN GOLD MINE TAILINGS SLURRY

SAFETY CONSIDERATIONS

Safe operation of the process was of paramount importance and was insured by the following steps.

1. Several safety features were built into the generator itself to maximize safe operation. These included separation of the mixer-reactor from the feed skid/control unit, and a programmable logic controller (PLC) to monitor and control the chemical feed rates. The PLC also controlled alarms and emergency shutdowns. Also, design features eliminated the possibility of Caro's acid back-flowing into the feed system.

2. A Hazardous Operation study was conducted on the Caro's acid generator and the proposed test area.

3. Bags of soda ash were provided to clean up any spills of sulfuric acid that might occur.

CYANIDE NOMENCLATURE

The term "WAD CN" as used in this paper, is the sum of true WAD (weakly acid dissociable) CN^- plus free CN^-. True WAD CN^- are weakly acid dissociable complexes of CN^-, e.g., the copper and nickel complexes. True WAD CN^- and free CN^- are specifically designated as such.

EXPERIMENTAL

OUTLINE OF TEST PLAN

Several runs were made on mill tailings slurry at initial WAD CN^- levels of 40-50 ppm and 100-120 ppm.
Mole ratios of H_2SO_4 to H_2O_2 for Caro's acid generation were either 2.0/1 or 2.5/1. Mole ratios of CA/CN^- were between 0.7/1 and 2.9/1.

ANALYTICAL

SAMPLING

Samples of feed slurry were taken at the head of the line and samples of treated slurry were taken as the slurry entered the tailings pond.

ANALYTICAL METHODS

WAD CN^- (free + WAD) was determined by the picric acid method and free cyanide by the silver nitrate titrimetric method using Rhodanine indicator. As a cross check, WAD CN^- and free CN^- were also determined by the LaChat Micro-Distillation method. The methods generally were in good agreement.

RESULTS AND DISCUSSION

1. Release Of HCN
     The atmospheric HCN detectors at the tailings treatment site registered non-detectable levels of HCN during most of the plant trial never exceeding 2 ppm HCN. The 2 ppm HCN maximum level was well within the 5 ppm limit at which point an alarm would be sounded. The low atmospheric HCN levels are attributed to the very rapid reaction of the Caro's acid with the cyanide.

2. Effect Of Feed CA/CN⁻ Mole Ratio
     How the dosage level of Caro's acid relative to the WAD CN⁻ in the feed affects CN⁻ detox is examined in this section.
     Figure 3 shows that a CA/CN⁻ mole ratio of 1/1 is sufficient to control effluent at 12-16 ppm WAD CN⁻, which is well below the 25 ppm target level.
     Figure 3 also indicates a refractory lower level of WAD CN⁻ which will not yield even to CA/CN⁻ ratios of greater than 2.5. This lower level is about 4 ppm WAD CN⁻ for the tailings tested. Results with other tailings slurries may result in lower or higher final WAD CN⁻ levels and will depend in large part on the type of WAD CN⁻ that are present.

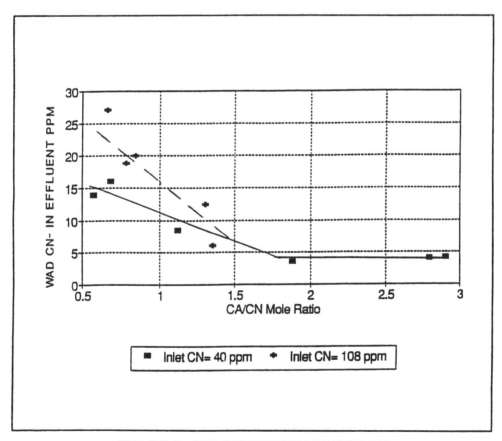

FIGURE 3 - CONCENTRATION OF WAD CN⁻
IN EFFLUENT VS CA/CN⁻ MOLE RATIO

3. Effect Of Operating Parameters On Effluent pH

Figure 4 shows the relation of effluent pH to CA/ CN⁻ mole ratios for 40 ppm and 108 ppm inlet WAD CN⁻. At a CA/CN⁻ mole ratio of 1/1 and a H_2SO_4/H_2O_2 mole ratio of 2.5/1, effluent pH will be 8.8 and 6.6 for inlet WAD CN⁻ levels of 40 and 108 ppm respectively. At 108 ppm inlet WAD CN⁻, an increase of about 0.6 pH units occurs by decreasing the H_2SO_4/H_2O_2 mole ratio from 2.5 to 2.0. The effect of lowering the H_2SO_4/H_2O_2 mole ratio was too small to be measured under plant conditions at 40 ppm inlet WAD CN⁻. Lowering the H_2SO_4/H_2O_2 mole ratio however, raises the adiabatic mix temperature. This may not be desirable since the higher temperature can lead to greater instability of the Caro's acid.

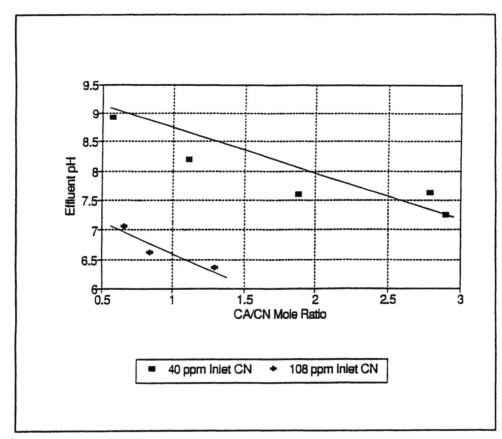

FIGURE 4 - EFFLUENT pH VS CA/CN⁻ MOLE RATIO

4. Free CN⁻ And True WAD CN⁻ Destruction

Data on the relative ease of destruction of free and true WAD CN⁻ is shown in Figures 5-6. It is more difficult to destroy true WAD CN⁻ than free CN⁻. For example, Figures 5 and 6 indicate that at an inlet CA/CN⁻ mole ratio of 1.0, about 90-95% of the free CN⁻, but only 20-40% of the WAD CN⁻ would be destroyed. However, by increasing the mole ratio of CA/CN⁻ to 1.8/1, the amount of true WAD CN⁻ destroyed increases to 90%.

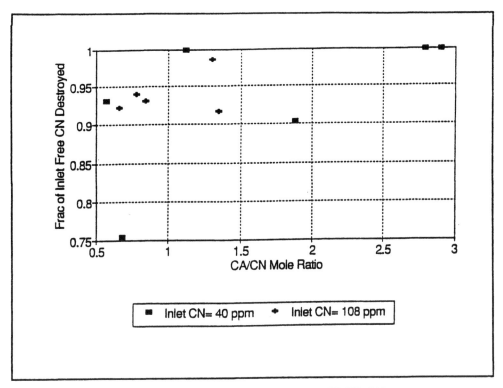

FIGURE 5 - FRACTION OF INLET FREE CN⁻
DESTROYED VS CA/CN⁻ MOLE RATIO

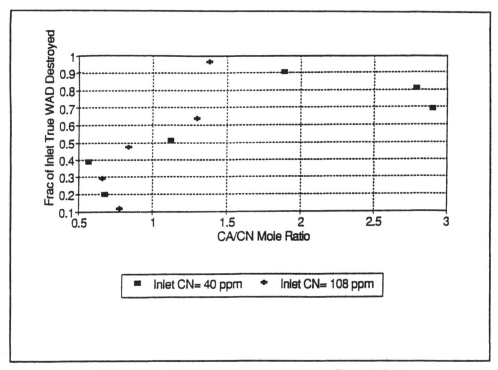

FIGURE 6 - FRACTION OF INLET TRUE WAD CN⁻
DESTROYED VS CA/CN⁻ MOLE RATIO

5. Economics

The raw material costs for Caro's acid is comprised of the cost of peroxide, sulfuric acid and added CaO (lime) for neutralization.

An estimated raw material cost breakdown for the Caro's acid is as follows.:

	% of Total Cost	
	w/CaO	w/o CaO
H_2O_2	57	67
H_2SO_4	29	33
CaO	14	0

For this plant trial, extra lime was not needed but in less buffered tailings slurries, extra lime might be required during the addition of the Caro's acid. In addition, since most mines recycle a portion of their tailings pond back to the mill, some neutralization will be required either during or after Caro's acid treatment. If the neutralization is conducted only on that portion of water reclaimed, then neutralization costs will be less than if the entire tailings slurry is neutralized.

The raw material cost to destroy one pound of CN^- is plotted, including CaO cost for complete neutralization of the acid in Figure 7. Figure 7 shows that the cost per pound of CN^- destroyed is independent of inlet CN level.

Figure 8 shows \$/lb of CN^- cost, excluding CaO. Thus if CaO is not needed, then for a 20 ppm effluent CN^- level, the cost without CaO is \$1.30 vs \$1.55/lb of destroyed CN^- with CaO.

Raw material costs on a \$/ton dry ore basis are shown in Figure 9. This basis will be dependent on the inlet CN^- loading and percent solids in the slurry. Figure 9 shows the total cost, including CaO for neutralization as a function of exit WAD CN for 40 and 108 ppm CN^- inlet loadings. Obviously, the \$/ton cost is higher for the higher inlet CN^- concentration.

Finally, the \$/T cost to achieve either 10 or 20 ppm CN^- in the effluent is plotted in Figure 10 over a range of inlet CN^- concentrations. It costs about \$0.10/ton in total raw material cost to bring inlet WAD CN^- from 40 ppm down to 20 ppm. This compares to \$0.32/ton using H_2O_2 alone. Thus Caro's acid has the potential to substantially reduce raw material costs for CN^- detoxification.

The capital cost for a Caro's acid generator is < \$200,000
The costs for the H_2O_2 and H_2SO_4 storage vessels are not included but would be significantly less than the investment in the Caro's acid unit.

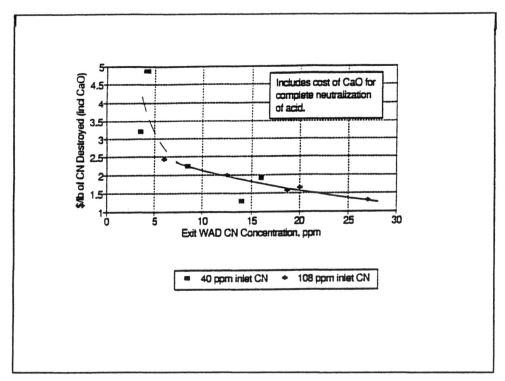

FIGURE 7 - COST OF CN⁻ DESTRUCTION VS CN⁻ EXIT CONC.

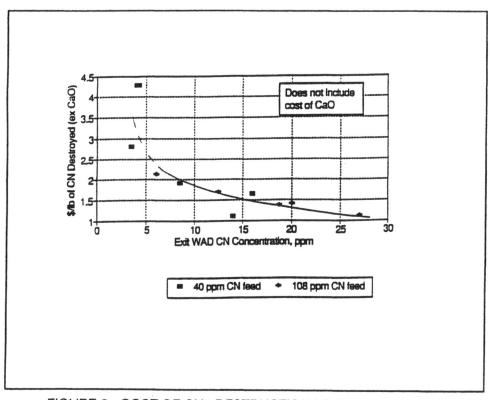

FIGURE 8 - COST OF CN⁻ DESTRUCTION VS CN⁻ EXIT CONC.

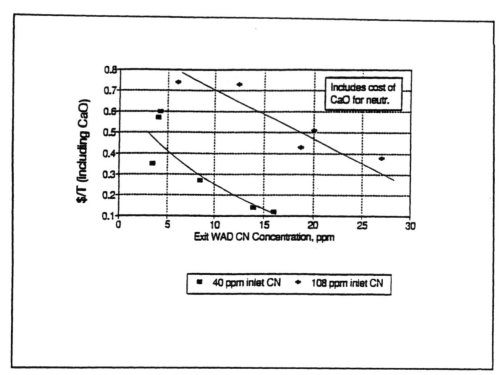

FIGURE 9 - COST OF CN⁻ DESTRUCTION VS CN⁻ EXIT CONC.

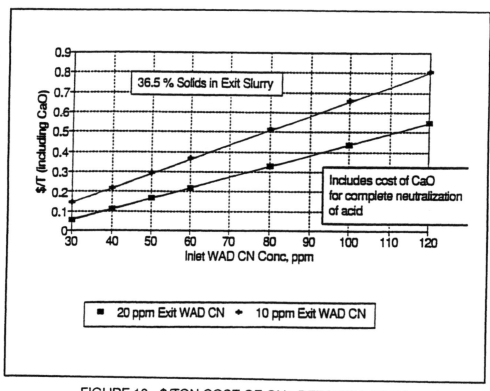

FIGURE 10 - $/TON COST OF CN⁻ DETOXIFICATION
WITH CARO'S ACID VS INLET WAD CN⁻ CONCENTRATION

CONCLUSIONS

1. Caro's acid (peroxymonosulfuric acid) was successfully field tested as a cyanide detoxification agent.

2. Substantial raw material savings and shortened cyanide detoxification times compared to using H_2O_2 alone were realized.

3. Atmospheric HCN levels at the tailings treatment site were at 0-2 ppm over the course of the plant trial even though extra alkali was not added.

4. Using a 1/1 mole ratio of Caro's acid/CN^-, and a 2/1 or 2.5/1 mole ratio of H_2SO_4/H_2O_2, WAD CN^- levels were reduced from 40 -108 ppm to below the target concentration of 20 ppm. By increasing the CA/CN^- mole ratio to 1.5/1, WAD CN^- levels were reduced to 4 ppm.

5. At a CA/CN^- mole ratio of 1/1, the effluent pH dropped from an initial pH of 10.5 to a pH of 8.8 and 6.6 for feed WAD CN^- levels of 40 and 108 ppm respectively.

6. The raw material costs for the detoxification were reduced from \$0.32/dry ton ore with H_2O_2 to \$0.10/dry ton ore with Caro's acid starting with a WAD CN^- concentration of 40 ppm.

REFERENCES

1. Castrantas, Harry M, and Michael R. Fagan U.S. Patent 5,137,642 "Detoxification of Aqueous Cyanide Solutions" Issued August 11, 1992

2. Betermier, Benedicte., Manual Alvarez, and Robert D. Norris U.S. Patent 4,446,029 "Destruction of Iron Cyanide Complexes" Issued May 1, 1984.

3. Schiller, Joseph E.,1983 "Removal of Cyanide and Metals From Mineral Processing Waste Waters" Bureau of Mines Report RI 8836 United States Department of the Interior, Washington DC.

4. Chiang John S. U.S. Patent 4,049,786 "Process of Preparing Peroxymonosulfate" Sept 20, 1977

MARTIN JEFF
ALASTAIR McNEILLIE
STEPHEN F. ROBINSON

Practical Application of Peroxygen Compounds for the Treatment of Cyanide Laden Wastewater in the Presence of Sulfur Containing Species

ABSTRACT

Wastewater containing cyanide in both simple and complexed forms is generated by a variety of industries including, for example, petrochemical refining and goldmining. These wastewaters often contain additional contaminating species such as sulfide or thiocyanate ions which render the effluents rather more complex. The potential application of chemical oxidation technology via treatment with peroxygen reagents such as hydrogen peroxide or Caro's acid (H_2SO_5) can be markedly affected by these additional contaminants. This paper describes the real effects observed and the resultant practical implications for treatment of two different types of cyanide laden industrial wastewater with peroxygen based systems when either sulfide or thiocyanate ions were additionally present.

INTRODUCTION

Application of hydrogen peroxide or peroxymonosulfuric acid (Caro's acid) for the chemical oxidation of cyanide containing wastewaters at alkaline pH has been previously described [1-3]. The representative reactions with either hydrogen peroxide or peroxymonosulfate are:

$$CN^- + H_2O_2 \longrightarrow CNO^- + H_2O \tag{1}$$

$$CN^- + HSO_5^- \longrightarrow CNO^- + HSO_4^- \tag{2}$$

$$CNO^- + 2H_2O \longrightarrow HCO_3^- + NH_3 \tag{3}$$

When contemplating the use of such oxidation reagents for removal of cyanide contamination, several factors need to be taken into consideration, including the availability of the reagents, ease of storage and handling, efficiency of cyanide removal with respect to treatment targets, potential interferences, and treatment economics.

Martin Jeff, Solvay Interox, 1230 Battleground Road, Deer Park, TX, 77536
Alastair McNeillie and Stephen F. Robinson, Solvay Interox, 3333 Richmond Avenue, Houston, TX, 77098

Actual industrial wastewater streams rarely contain only one single pollutant, rather they are frequently complex mixtures of a range of contaminants, many of which exert there own chemical oxygen demand. Such additional compounds to the pollutant of primary interest may react readily with peroxygen reagents and markedly affect the outcome of the treatment scheme.

Recently, two examples of such co-contaminant effects were experienced by Solvay Interox when evaluating the use of peroxygen reagents for detoxification of cyanide containing wastewaters. Both involved the presence of inorganic sulfur species and are presented in detail in the following case studies.

CASE STUDY 1 - SULFIDE AND CYANIDE

BACKGROUND

A Southern United States petroleum refining operation routinely generates wastewater containing both sulfide and cyanide. This is normally treated with ferrous sulfate to reduce the cyanide level via precipitation of a ferro/ferricyanide compound before the wastewater passes on to a biological treatment unit. The company was interested in the potential use of H_2O_2 for oxidative pretreatment of the wastewater to remove cyanide. A laboratory feasibility study was initially conducted on wastewater samples to determine optimum treatment conditions. The initial wastewater was characterized as follows:

Weak Acid Dissociable Cyanide (CN_{WAD}) = 859 mg/L

Total Inorganic sulfide = 1000 mg/L

pH = 6.9

The presence of a significant amount of sulfide was expected to markedly affect the efficiency and outcome of treatment with H_2O_2. Hydrogen peroxide rapidly oxidizes sulfide to colloidal sulfur (S^0) at acidic pH and to sulfate (SO_4^{2-}) at alkaline pH according to reactions (4) and (5), respectively [4,5].

$$\text{Acidic pH} \quad H_2S + H_2O_2 \text{-----}> S^0 + 2H_2O \qquad (4)$$

$$\text{Alkaline pH} \quad S^{2-} + 4H_2O_2 \text{-----}> SO_4^{2-} + 4H_2O \qquad (5)$$

At neutral pH, both colloidal sulfur and sulfate are produced according to reactions (6) and (7). Periodic oscillation and bistability of redox potential and pH have actually been demonstrated at near neutral pH as transient yellow polysulfides are first formed followed by swings between colorless sulfate and white colloidal sulfur [6].

$$HS^- + H_2O_2 + H^+ \dashrightarrow S^0 + 2H_2O \qquad (6)$$

Neutral pH

$$HS^- + 4H_2O_2 \dashrightarrow SO_4^{2-} + 4H_2O + H^+ \qquad (7)$$

Consequently, consumption of H_2O_2 by sulfide is directly proportional to pH, increasing from 1 lb H_2O_2/lb H_2S at pH <6 to 4.25 lbs H_2O_2/lbs S^{2-} at pH >10. Because of this varying demand for H_2O_2 with pH, it was decided to examine treatment of the cyanide wastewater at both pH 9 and pH 6. Removal of cyanide with H_2O_2 is typically carried out in the pH range 9.5 - 11.0 but the anticipated higher demand for H_2O_2 in this case was considered a potential economic barrier to effective treatment.

METHODOLOGY

Feasibility studies were conducted on 500 mL samples of wastewater adjusted to the desired pH just prior to treatment with either dilute sulfuric acid or dilute sodium hydroxide solution. Reactions were carried out at 32°C for a total of two hours. Samples were dosed with 50% hydrogen peroxide at a level of either 3:1 or 5:1 H_2O_2:CN_{WAD} on a weight/weight basis. This corresponded to an actual dose of either 2576 mg/L or 4294 mg/L H_2O_2 (100% basis) for wastewater initially containing 859 mg/L CN_{WAD}. Dosing was based on cyanide since this was the primary pollutant of interest to the enduser.

RESULTS AND DISCUSSION

The effects of the different treatments on cyanide and residual H_2O_2 levels are summarized in Figure 1. Additionally, all treatments resulted in complete removal of sulfide.

For reactions conducted at pH 6, this efficient removal of sulfide was as expected based on the levels of H_2O_2 dosed [effectively 2.6:1 or 4.3:1 H_2O_2:S^{2-} (w/w)]. It can be seen that excellent reduction of cyanide levels was also observed at pH 6 indicating the potential for an efficient H_2O_2 treatment system on a larger scale. Significant levels of residual hydrogen peroxide also remained after two hours and suggests that larger reaction times may result in further lowering of cyanide concentrations.

Removal of all sulfide at pH 9 with a dose of only ca. 2600 mg/L H_2O_2 vs. 1000 mg/L S^{2-} was at first somewhat surprising, especially when considered in conjunction with the large reduction in cyanide that was also obtained. Even treatment with the higher dose of H_2O_2 at pH 9 provided results that were better than initially anticipated since both sulfide and cyanide were very effectively removed.

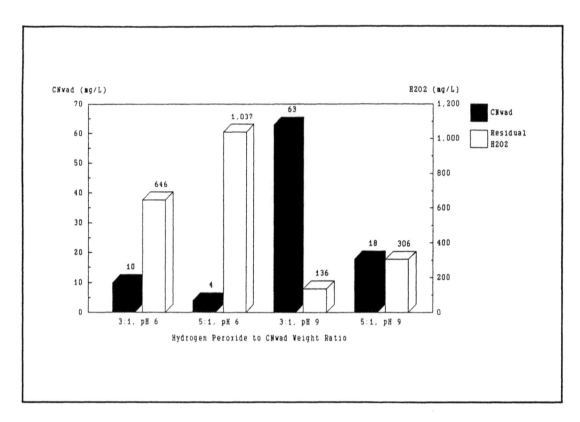

Figure 1. Effect of H_2O_2 on Cyanide/Sulfide Wastewater

A possible explanation of the observed behavior comes to light when the corresponding production of thiocyanate in these systems is considered. The levels of thiocyanate attained for the various treatments applied are indicated in Figure 2.

For reactions at pH 9, significant amounts of thiocyanate were generated. It is suggested that this species arises as a result of reaction between cyanide and some of the intermediate sulfide oxidation products such as polysulfide or thiosulfate [7,8]. It is interesting to note that only small amounts of thiocyanate were produced in the treatments carried out at pH 6. This could be a consequence of either the different mechanism and extent of sulfide oxidation at pH 6 vs. pH 9 or greater direct oxidation of cyanide with H_2O_2.

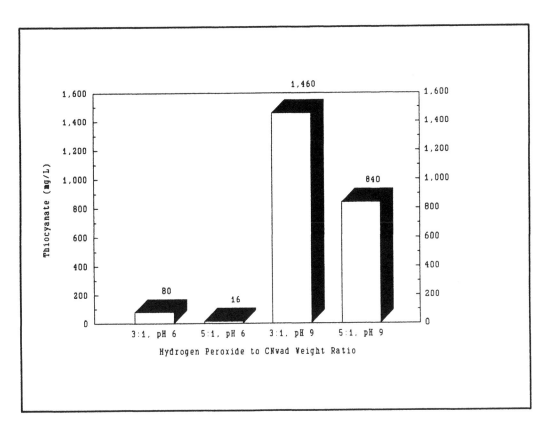

Figure 2. Generation of SCN via Oxidation of Cyanide/Sulfide

After reviewing the results of the feasibility study, the customer decided to proceed with field trials based on treatment with H_2O_2 at pH 6 rather than pH 9. The lower cyanide levels attained, combined with little thiocyanate generation and the need for less pH adjustment were all attractive benefits. Furthermore, additional toxicity testing using a Microbics Microtox® system had indicated excellent reduction in toxicity after treatment at pH 6. Treated wastewater had an EC_{50} = 100% compared to EC_{50} = 0.5% for the untreated initial sample.

Field implementation did not achieve quite as high a level of cyanide removal as observed in the laboratory. Treatment with 5-6 parts H_2O_2 per part CN_{WAD} resulted in a removal of CN_{WAD} of 72-74%. However, the exact levels of sulfide in the incoming wastewater were not carefully monitored during the trials and a higher ratio of sulfide to cyanide may have affected the efficiency of treatment. Nevertheless, sulfide was completely eliminated upon treatment but the significant amounts of elemental sulfur produced gave rise to handling problems such as blocked lines and pumps. Full-scale implementation of the technology would require a system for handling and/or recovery of the sulfur produced.

CASE STUDY 2 - CYANIDE AND THIOCYANATE

BACKGROUND

A North American goldmining operation has tailings ponds containing both free and complexed cyanide species as well as copper and iron metal ions. The majority of this water is recycled in the plant but periodically, discharge of overflow water is required as the ponds approach capacity. Previously, they have employed a hydrogen peroxide treatment system at alkaline pH for oxidation of cyanide prior to discharge. The discharge limits for cyanide, copper, and iron in this case are very low and are as indicated below:

$$\text{Total Cyanide } (CN_{TOT}) \quad = 1.5 \text{ mg/L max.}$$

$$\text{WAD Cyanide } (CN_{WAD}) \quad = 0.1 \text{ mg/L max.}$$

$$\text{Copper} \quad = 0.15 \text{ mg/L max.}$$

$$\text{Iron} \quad = 3 \text{ mg/L max.}$$

Utilizing the existing hydrogen peroxide treatment system, these discharge limits have only been approached by employing a $H_2O_2:CN_{TOT}$ mole ratio of 15:1. The client was very interested in potential improvements in a peroxygen treatment process and anything that would increase the efficiency of the oxidant towards cyanide destruction.

Caro's acid (active species peroxymonosulfuric acid, H_2SO_5) has been shown to be particularly effective for oxidation of cyanide species [1,3,9]. The reaction proceeds over a wide pH range without a catalyst but is typically kept alkaline (pH 9.5 - 10.5) to prevent formation of hydrogen cyanide. Typically, 1.5 - 1.8 moles of HSO_5^- are required to oxidize one mole of CN^-. Because of the greater efficiency of Caro's acid versus H_2O_2 for destruction of cyanide, laboratory trails were conducted on the clients wastewater in an effort to demonstrate superior treatment performance.

Feasibility studies were conducted on wastewater samples from two separate tailings ponds which were characterized as indicated in Table I. The presence of significant amounts of thiocyanate, however, was expected to impact the efficiency of treatment since peroxymonosulfate may be consumed by interaction with thiocyanate according to reactions (8) and (9), which can be simplified to the overall reaction (10) [10-12].

$$SCN^- + 3HSO_5^- + H_2O \longrightarrow 4SO_4^{2-} + 4H^+ + HCN \qquad (8)$$

$$HSO_5^- + HCN \longrightarrow SO_4^{2-} + H^+ + OCN^- \qquad (9)$$

$$SCN^- + 4HSO_5^- + H_2O \longrightarrow 5SO_4^{2-} + 6H^+ + OCN^- \qquad (10)$$

TABLE I - INITIAL WASTEWATER FOR TREATMENT

Parameter	Pond #1	Pond #2
pH	7.5	11.9
CN_{TOT} (mg/L)	99	390
CN_{WAD} (mg/L)	98	355
CN_{FREE} (mg/L)	30	334
SCN (mg/L)	240	370
Cu (mg/L)	103	171
Fe (mg/L)	5.1	12

METHODOLOGY

A laboratory sample of Caro's acid was prepared by the controlled addition of 50% H_2O_2 to 98% sulfuric acid with cooling and stirring. A molar ratio of H_2SO_4:H_2O_2 of 2:1 was utilized which resulted in an initial solution of Caro's acid with a concentration of H_2SO_5 of ca. 28% w/w. A portion of this "as-made" Caro's acid was then diluted with the corresponding amount of demineralized water required to produce a concentration of ca. 10% w/w. The diluted Caro's acid thus obtained was analyzed for H_2O_2 and H_2SO_5 content just prior to its addition to wastewater samples and found to contain 2.05% and 7.9%, respectively.

Reactions were conducted by dosing a diluted solution of H_2O_2 (ca. 5% w/w), or the diluted Caro's acid solution, to one (1) liter samples of wastewater from Pond #1 or Pond #2. The initial temperature of all solutions was ca. 20 °C and no significant exotherms were observed in any case upon dosing of the peroxygen reagent.

For treatment of Pond #1 samples with H_2O_2, the pH was first raised to ca. pH 10.1 (from pH 7.25) by addition of 10% NaOH solution. Dosing was complete in 2-3 minutes, and the pH of the final solution was then adjusted to ca. pH 10.2. For Pond #2 samples, the initial pH was already at pH 11.9 so that H_2O_2 was dosed without prior adjustment. The treated solution obtained was, however, adjusted to ca. pH 10.2 after H_2O_2 addition was complete.

For treatment of wastewaters with Caro's acid, manual control of pH during dosing of the acidic reagent was employed. A solution of either 10% or 25% NaOH was utilized, dependant upon the total amount of Caro's acid reagent to be added. This crude pH control resulted in the pH never dropping below pH 9.5 and for the majority of the dosing period the pH was between pH 10.5 - 11.0.

After dosing of the relevant peroxygen reagent, reactions were allowed to stand at room temperature for 2-3 hours after which time the presence of residual available oxygen was checked. If necessary, 3 mLs of a 4% As_2O_3 solution were added to kill any remaining residual peroxide. Precipitates were removed by vacuum filtration and the filtrates obtained preserved and stored for subsequent analysis.

RESULTS AND DISCUSSION

The effects of treatment of the wastewater samples with either hydrogen peroxide or Caro's acid are presented in Figure 3 (Pond #1) and Figure 4 (Pond #2). The data indicates that Caro's acid was indeed more efficient than H_2O_2 for destruction of cyanide. Lower residual cyanide levels were obtained with smaller molar doses of Caro's acid versus H_2O_2. For Pond #1, the target levels of CN_{WAD} and CN_{TOT} were reached or surpassed (0.1 mg/L and 0.28 mg/L, respectively) with a dose of Caro's acid of only 2.4:1 H_2SO_5:CN_{TOT}. Addition of further amounts of Caro's acid did not, however, improve upon this good result.

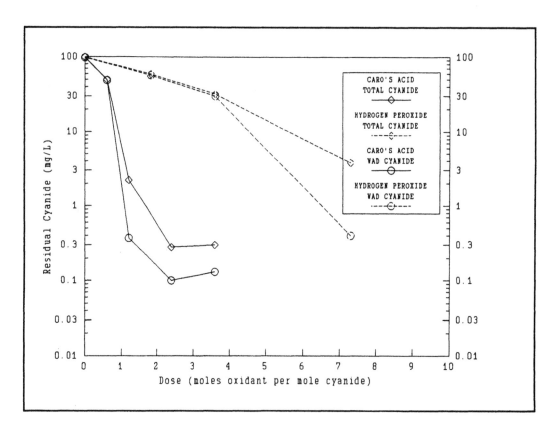

Figure 3. Cyanide Levels Pond #1

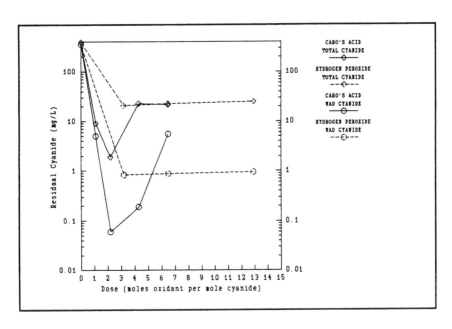

Figure 4. Cyanide Levels Pond #2

For Pond #2, the target cyanide levels were again very closely approached (CN_{TOT} = 1.9 mg/L and CN_{WAD} = 0.06 mg/L) when a mole ratio of H_2SO_5:CN_{TOT} of 2.2:1 was applied. In contrast, treatment with high levels of H_2O_2 (12.9:1 molar ratio) failed to achieve the target levels. However, despite the good results obtained at relatively low doses of Caro's acid, the residual cyanide levels increased as the applied dose of Caro's acid was further increased, resulting in poor overall treatment w.r.t. discharge limits. This observed phenomenon is believed to be due to the increasing role of thiocyanate oxidation as the amount of H_2SO_5 applied increases. The corresponding levels of thiocyanate observed after treatment are presented in Figure 5.

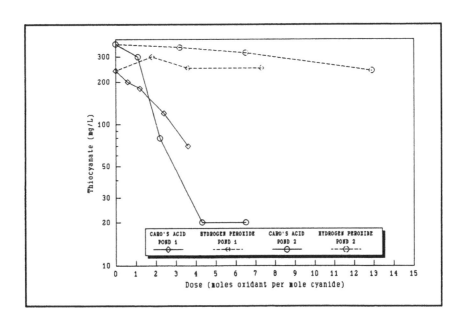

Figure 5. Thiocyanate Levels - Ponds #1 & #2

Treatment of Pond #1 or #2 with H_2O_2 showed little effect on the levels of thiocyanate, even at high mole ratios. In contrast, the dramatic removal of SCN with H_2SO_5 can be clearly seen. The increase in cyanide levels observed for Pond #2 as thiocyanate continues to decrease with higher doses of Caro's acid is possibly due to regeneration of cyanide according to reaction (8) described above. It is of further note that for the two larger doses of H_2SO_5:CN_{TOT} of 4.3 and 6.5:1 applied to Pond #2, the effective molar ratio of H_2SO_5 to the sum of (CN_{TOT} + 4SCN) was 1.6 and 2.4:1, respectively. Presumably, an even greater amount of H_2SO_5 would be required to ultimately reduce the cyanide levels again after all the thiocyanate was consumed. Such a dose of Caro's acid would not, however, be cost effective and would also add large amounts of sulfate by-product to the wastewater which may not be desirable.

It is also important to evaluate the effectiveness of the peroxygen treatments in terms of copper and iron removal since these two species are also subject to low discharge limits. The results obtained for copper and iron are presented in Figures 6 and 7, respectively. Overall, neither hydrogen peroxide or Caro's acid were successful in lowering the copper content of Pond #1 or #2 to the desired level. Only the high dose of Caro's acid approached the target for copper when applied to Pond #2. As we have already seen, however, this treatment was not effective due to the significant cyanide concentrations that remained under such conditions.

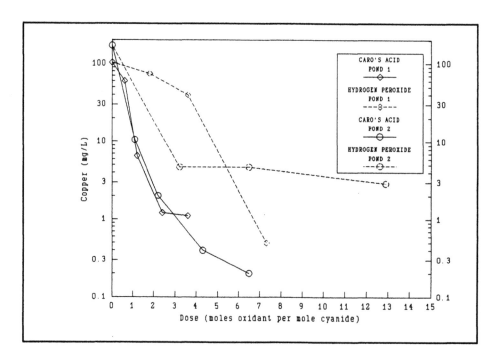

Figure 6. Copper Levels - Ponds #1 & #2

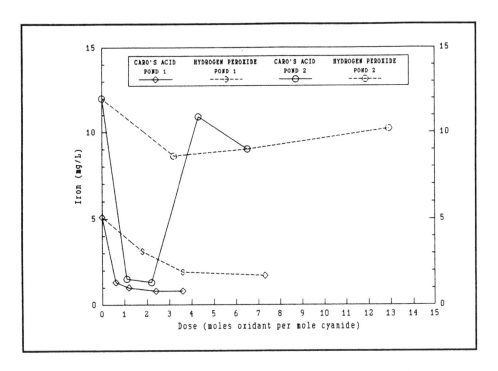

Figure 7. Iron Levels - Ponds #1 & #2

Removal of iron from the wastewaters was more readily achieved than the removal of copper. Caro's acid was effective for both Ponds when dosed in the range 2.2 - 2.4:1 H_2SO_5:CN_{TOT}. Higher doses of Caro's acid were ineffective for iron removal when applied to Pond #2. This behavior correlates with the postulated reformation of cyanide as a result of extensive thiocyanate oxidation. Hydrogen peroxide treatments were only effective for removal of iron from Pond #1 and then only at the higher doses applied.

In summary, it would appear that Caro's acid is much more efficient than hydrogen peroxide for detoxification of the gold mine effluent waters examined here. However, because of the presence of relatively high levels of thiocyanate, an optimum, relatively low dose of H_2SO_5 was observed to be effective. With such a dose, three of the four critical discharge parameters could be very closely approached if not surpassed. Unfortunately, under such conditions, the required copper discharge limits could not be attained.

CONCLUSIONS

The presence of more than one contaminant in a wastewater stream can have a dramatic affect on the practical outcome of a treatment based on chemical oxidation. The examples described in this paper serve to remind us of the importance of evaluating the "total" system before an effective and practical pollution control scheme can be implemented. In some cases, positive synergy may be experienced such as was the case with alkaline treatment of cyanide wastewater in the presence of sulfide. In contrast, application of a stronger chemical oxidant may give rise to additional complications such as observed with Caro's acid treatment of cyanide/thiocyanate wastes. The efficacy, practicality, specificity, and total cost must be weighed before finalizing the selection of a particular treatment scheme.

REFERENCES

1. Gregor, K. H., 1991. "Cyanide Detoxification with Peroxygens," in Chemical Oxidation:Proceedings of the 1st International Symposium, W. W. Eckenfelder, A. R. Bowers, and J. A. Roth, ed. Lancaster, PA:Technomic Publishing Company, Inc., pp. 96-103.

2. Kurek, P. R., R. R. Frame, T. N. Kalnes, and M. D. Moser, 1993. "Oxidative Removal of Cyanide from Aqueous Streams," in Chemical Oxidation:Proceedings of the 3rd International Symposium, W. W. Eckenfelder, A. R. Bowers, and J. A. Roth, ed. Lancaster, PA:Technomic Publishing Company, Inc., pp. 28-33.

3. Castrantas, H. M., J. L. Manganaro, C. W. Rautiola, and J. Carmichael, 1993. "The Caro's Acid Detoxification of Cyanide in a Gold Mine Tailings Pond - A Plant Demonstration," Randol Gold Forum, Beaver Creek, CO, pp. 337-342.

4. Hoffman, M. R., 1977. "Kinetics and Mechanism of Oxidation of Hydrogen Sulfide by Hydrogen Peroxide in Acidic Solution." Environ. Sci. & Technol., 11(1):61-66.

5. Jolley, R. A. and C. F. Forster, 1985. "The Kinetics of Sulfide Oxidation." Environ. Tech. Letters, 6:1-10.

6. Orban, M., and I. R. Epstein, 1985. "A New Halogen-Free Chemical Oscillator:The Reactor Between Sulfide Ion and Hydrogen Peroxide in a CSTR." J. Amer. Chem. Soc., 107(8):2302-2305.

7. Scott, J. S., and J. C. Ingles, 1981. "Removal of Cyanide from Gold Mill Effluents," Canadian Mineral Processors 13th Annual Meeting, Ottawa, ONT.

8. Schiller, J. E., 1983. "Removal of Cyanide and Metals from Mineral Processing Waste Waters," USDI Bureau of Mines, Report # 1983-705-020/97, Washington, DC.

9. Interox, October 1987. "Cyanide Waste Detoxification." Effluent & Water Treatment Journal, pp. 42-48.

10. Smith, R. H., and I. R. Wilson, 1966. "The Mechanism of the Oxidation of Thiocyanate Ion by Peroxomonosulfate in Aqueous Solution, Part I." Aust. J. Chem., 19:1357-63.

11. Smith, R. H., and I. R. Wilson, 1966. "The Mechanism of the Oxidation of Thiocyanate Ion by Peroxomonosulfate in Aqueous Solution, Part II." Aust. J. Chem., 19:1365-75.

12. Smith, R. H., and I. R. Wilson, 1967. "The Mechanism of the Oxidation of Thiocyanate Ion by Peroxomonosulfate in Aqueous Solution, Part III." Aust. J. Chem., 20:1353-66.

VANEATON PRICE III
DONALD L. MICHELSEN
RILEY T. CHAN
JEFF W. BALKO

Continuous Color Removal from Concentrated Dye Waste Discharges Using Reducing and Oxidizing Chemicals—A Pilot Plant Study

ABSTRACT

The purpose of this research was to design, fabricate and test a 1 liter per minute pilot plant with a cascading sequence of continuously stirred tank reactors. The object of the research was to chemically decolorize selected reactive-dye bath concentrates resulting from exhaustive dyeing, and to remove metals and DOC using Fenton's Reagent or the reductive chemicals, thiourea dioxide and sodium hydrosulfite. For the Fenton's Reagent studies, ferrous sulfate was premixed with the dye waste concentrate before overflowing to the first reactor.

A feedback control system based on color remaining in the discharge was used to regulate reactants added. Transmittance was measured at several wavelengths (590, 540, and 438 nm) and the American Dye Manufacturers Institute (ADMI) value calculated. The results demonstrated that ADMI measurements could not be made on dark solutions (over 3000 ADMI) in the pilot plant and, typically, one wavelength was used for control. DOC removal was used as a means of determining the biological activity in aerated reactors following color removal.

The initial pilot plant studies were conducted using Navy 106 jet-dye waste. Reductive pretreatment with thiourea dioxide resulted in 92.2% color removal with color returning upon aeration for an overall color removal of 76.6%. Oxidative pretreatment with Fenton's chemistry resulted in 98.8% color removal with overall color removal after aerobic treatment at 96.8%. Dissolved Organic Carbon (DOC) removal in aerobic treatment improved with oxidative pretreatment relative to reductive pretreatment on Navy 106 jet-dye concentrate.

On site operation of the pilot plant on other dye wastes showed color removals above 95% and DOC removals of 38% and 19% for an azo-based red dye waste concentrate and a copper-phthalocyanine-based dye, Ming Jade, respectively. The soluble copper concentration in the Ming Jade was decreased from 19.2 ppm to 4.5 ppm. This corresponded to a

Vaneaton Price III, Donald L. Michelsen, Riley T. Chan and Jeff W. Balko, Virginia Polytechnic Institute and State University, Blacksburg, Virginia 24061, USA

3-fold increase in suspended solids from 0.575 g/L to 1.505 g/L.

The results showed that continuous oxidative pretreatment with a 15-minute residence time was controllable and more effective than reductive treatment for color removal. Oxidative pretreatment also decreased the soluble copper concentration in a copper containing waste water, and did not hinder biological activity.

INTRODUCTION

The fate of dyestuffs in waste streams resulting from textile dyeing has long been a concern for reasons of esthetics and toxicity in the environment. Today color removal is generally carried out by municipal wastewater treatment facilities. Fiber reactive dyes contribute greatly to the total color discharged to these Publicly Owned Treatment Works (POTWs) and, because of their high water-solubility, are not easily treated by typical activated-sludge treatment facilities (Pagga and Brown, 1986). Any color removal that is achieved in an aeration tank is more or less accidental and is the result of adsorption of color onto sludge which is removed by flocculation (Meyer, 1981).

Because of this ineffectiveness, color is typically removed using polymer sorbents at a very substantial cost to those companies discharging color to the POTWs, or by treatment with chlorine which must be followed by dechlorination, which can lead to a more toxic discharge. Furthermore, the dye-saturated polymers must be sent to landfills or incinerated for final disposal, creating an alternative demand on the environment. Also, increased pressure from regulatory agencies to reduce the final color value of discharges to waters of the United States has intensified the desire to find alternative methods for more effectively treating color in dye waste streams.

Previous research has shown that both oxidative (Powell, et al., 1992), and reductive (McCurdy, et al., 1991) chemistry in the form of pre-treatment can be very effective for color removal in streams containing fiber-reactive azo and disazo dyes. Reductive pretreatment followed by biological treatment in sequenced batch reactors, while successfully removing color, was found by McCurdy to result in residuals that limited the total percent removal of color, COD, BOD, and TOC in aerobic treatment. Oxidative schemes, however, were shown to neither enhance nor hinder DOC removal in biological reactors and pose no threat to activated-sludge wastewater treatment plants (Powell, et al., 1992).

The idea for the continuous pre-treatment of textile

dye waste was presented on April 5, 1991 (Mann and Woodby, 1991). The objective of the project was to "build, test, and use a portable, semi-automated, continuous flow pilot plant to chemically treat cotton dye wash streams..." The pilot plant was designed to use three source wavelengths of light to measure transmittance that could also be converted to an American Dye Manufacturers Institute (ADMI) color value (a value used to regulate local discharges). The ADMI value was then used for feedback control of the amount of pre-treatment chemical used in the reactor. The pilot plant was operated with a dye waste input rate of 1 L/min.

The pilot plant was designed to use strong reducing agents such as sodium hydrosulfite or thiourea dioxide, or oxidative chemistry in the form of Fenton's Reagent (ferrous iron and hydrogen peroxide) to cleave the azo bonds of reactive dyes. The system has the capacity to continuously measure pH or redox potential from a probe in the recycle loop.

The dye wastes tested were the Navy 106 jet-dye concentrate and Navy 106 slack washer waste from Tultex (Martinsville, Virginia), which contained three Remazol dyes (Hoechst Celanese Corporation). The three dyes were Remazol Black B (reactive Black 5), Remazol Red RB (Reactive Red 198), and Remazol Golden Yellow 3RA (reactive Orange 16). This formulation was chosen because of the high frequency of its use and its high contribution to overall color discharged by the plant. The feed tank to the reactor was maintained at 55 to 65 °C, to mimic dye-bath and rinse-effluent conditions in the textile mill.

The object of this research was (a) to operate and optimize the pilot plant for the pre-treatment of the concentrated jet-dye waste stream, and (b) to find those conditions at which maximum color removal could be attained with ease of control. After long-term, steady-state runs in the lab with stored dye waste, the pilot plant was transported to the Tultex mill and operated with fresh concentrated jet-dye waste, including Navy 106, an azo-based red dye, and a copper phthalocyanine based dye, Ming Jade.

Finally, the resulting waste streams from the chemical pre-treatment of Navy 106 jet-dye concentrate were aerobically digested in continuous flow, lab-scale biological reactors. This was done to ensure that no ill effects would occur to municipal treatment systems downstream from the site of pre-treatment.

LITERATURE REVIEW

Chemical reduction has been proven to be very successful for color removal of the dyes used in the Navy 106 formulation (Michelsen, et al., 1991). In chemical reduction, the nitrogen-nitrogen (azo linkage) double bond is broken, destroying the chromophore of the system. When azo dye compounds are chemically reduced, the result is the regeneration of the aromatic amines from which the dyes were originally manufactured (Weber and Wolfe, 1987). The main dye constituent in the Navy 106 formula (Reactive Black 5), for example, reduces to a substituted aniline (p-(2-hydroxyethylsulfone)-aniline), and a substituted naphthalene (8-hydroxy-1,2,7-triamino-3,6-naphthalene disulfonic acid) (Bell, et al., 1992).

Because of their ease of handling, the reduction chemicals chosen for use in the pilot plant were thiourea dioxide (formamidine sulfinic acid, FAS) and sodium hydrosulfite (sodium dithionite). According to Michelsen, et al., "thiourea dioxide performed the best by requiring the least amount of reagent for the greatest color [removal]." Also, "more color removal per gram of each reducing chemical was observed for the more highly colored jet discharges." It is important to note that the above results were obtained from bench-scale batch reactions and not on a continuous basis. The optimum color removal obtained using Navy 106 jet-dye concentrate was with 200 ppm and 500 ppm of thiourea dioxide and sodium hydrosulfite, respectively, as shown in figure 1.

There are many known oxidants that will decolorize azo dye wastes, including chlorine, ozone, and hydrogen peroxide. Chlorine is reported to produce potentially toxic organic by-products (Tincher, 1991). Ozone was found by Powell, et al., to be very effective for wastewater streams that contain significant amounts of non-colored organic matter; however, he notes that streams in which most of the organics result from dyestuffs (e.g. Navy 106 jet-dye concentrate) effective decolorization could be achieved with hydrogen peroxide. Because of its availability and ease of handling, Fenton's reagent was chosen to oxidize dye waste in the pilot plant. Fenton's reagent uses ferrous iron in the form of iron salts to catalyze the reduction of hydrogen peroxide to an hydroxide ion and an hydroxyl radical, as shown in equation 1. The hydroxyl radicals become the oxidizing species of Fenton's reagent, and the ferrous iron

$$H_2O_2 + Fe^{2+} \rightarrow \bullet OH + OH^- + Fe^{3+}$$ **Eq. 1**

is reduced to ferric iron. The hydroxyl radical is an extremely powerful oxidant that reacts nonspecifically with organic compounds. This "broadcast" oxidative effect of Fenton's reagent makes it more effective on waste streams that contain concentrated dye waste with minimal auxiliary organics. The classical Fenton's reagent chemistry is performed in acidic media with relatively high hydrogen-peroxide-to-ferrous-iron ratios to prevent the precipitation of the iron salts into iron hydroxides.Another advantage of

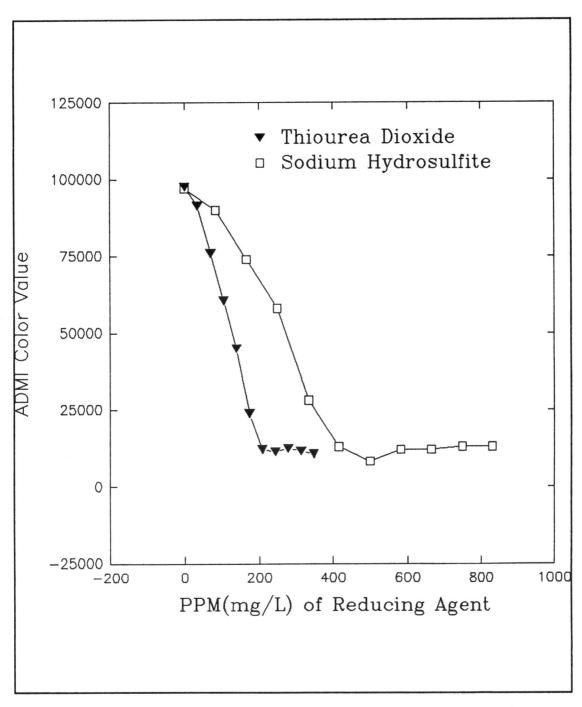

Figure 1 Reduction of Navy 106 jet-dye concentrate with varying concentrations of Thiourea Dioxide and Sodium Hydrosulfite (Michelsen, _et al_., 1991). Results shown were from bench scale batch reactions at pH 11.8 and 70°C.

Fenton's reagent technology is that it has been shown to effectively reduce the toxicity in effluents prior to aerobic treatment (Linneman, et al., 1991).

In bench-scale batch reactions, Fenton's reagent decreased the ADMI color value to very low levels (from 50,000 to 60,000 ADMI to below 1000 ADMI for Navy 106 jet-dye concentrate effluent), (Powell, et al., 1992). According to Powell's research, the ratio of hydrogen peroxide to ferrous iron is important, with lower ratios (higher iron concentrations) having lower final color values. The optimum dose for color removal was found to be 1000 ppm hydrogen peroxide in a 13.6-to-1 ratio with ferrous iron. Powell also noted that lower iron concentrations (20:1 H_2O_2:Fe^{3+} instead of 10:1 H_2O_2:Fe^{3+}) produced fewer solids. Solids are an issue in the feasibility of using transmittance for feedback control of the pilot plant because of the turbidity they cause and thus the possible interference with optical transmittance.

EXPERIMENTAL DESIGN

All experimentation for the pilot plant was carried out using dye waste resulting from the dyeing of the cotton portion of a 50/50 cotton/polyester blend. The dye formulation treated was the Navy 106 color used at Tultex in Martinsville, Virginia to dye cotton/polyester fleece. The cotton portion of the blend was dyed using Remazol dyes including Reactive Black 5, the single most used dye in the plant. The cotton can be dyed either by cold pad batch method or directly in the jet-dye machine used for polyester dyeing.

Fabric dyed by cold pad batch method is subsequently washed to remove excess dyes and chemicals. This results in the slack washer waste stream - the initial dye waste stream tested in the pilot plant. This stream is relatively dilute and typically has a color value between 2000 and 5000 ADMI. This stream has a high pH (>12) because of high caustic concentrations. It also contains surfactants, defoamers, and other dyeing assist chemicals.

The other waste stream tested results from jet-dyeing the cotton portion of the blend. This stream is simply the leftover dye after dyeing is complete. Jet-dye waste typically has a pH of approximately 11 and has a very high chloride concentration. The color of the jet-dye waste stream is typically between 55,000 and 75,000 ADMI.

For initial trials in the pilot plant to test the control scheme, slack washer waste was used. However, because jet-dye waste is a much greater contributor to the total color in the plant's effluent, all subsequent work was done on the jet-dye waste stream. Jet-dye waste also has high dye concentration with minimum auxiliary organics which makes it more attractive for treatment with Fenton's chemistry.

The pilot plant was operated using two different designs. The first design was used for reductive pretreatment and the second was used with oxidative (Fenton's) chemistry. Transformation between the two schemes was easily managed.

The design for reductive pretreatment (shown in figure 2) consisted of a 5 gallon bucket equipped with hydrostatic overflow that maintained a constant volume of 16 L. The recycle and feed were pumped using a masterflex peristaltic pump at 1.1 L/min which gave a residence time of 14.5 minutes. The feed was maintained at 60 °C by an electric heater and the reactor was continuously stirred by a three inch diameter prop driven by an electric motor.

The design used for oxidative pretreatment had a mix tank and three cascading reactors. The mix tank was used add and mix the ferrous iron solution and the dye waste feed and had a volume of 2 L. The first reactor was the main reacting vessel. It was used for the addition of the hydrogen peroxide solution, it had a volume of 10 L. The second reactor was maintained at 6 L and was used as a source and return for the recycle loop. Again, a masterflex peristaltic pump was used to feed dye waste to the pilot plant at 1.1 L/min, giving a residence time of 16.4 minutes. A schematic is shown in figure 3.

Both pilot plant schemes had a silicon glass cell located in the recycle loop (Figures 2 and 3). The photodiode receptors were covered with interference filters of nominal values of 589, 540, and 436 nm. A fourth photodiode, which was connected directly to the light source by a glass light pipe, was used for proportional feedback control of the light intensity.

Each of the three filtered photodiode detectors produces a current which is sent to separate current to voltage converters. The voltage is filtered and amplified to a of 0 to 5 V. An analog to digital converter is then used to convert the voltages to three 12-bit binary codes which are converted to ASCII by an IBM PC/XT which is running a compiled Quick BASIC program. The program converts the three signals to percent transmittance using a model taken from data for solutions measured both by the photodiodes and a Bausch & Lomb Spectronic 20 spectrophotometer (Balko, 1991). The transmittance values are used to calculate Munsell values which are used in the Adams-Nickerson difference equation (APHA, 1989) to determine the control variable (ADMI).

The program uses PID (proportional, integral, derivative) control to compare the calculated ADMI value to the set point and determine the response which is sent to the reagent pump. A digital to analog converter is used to

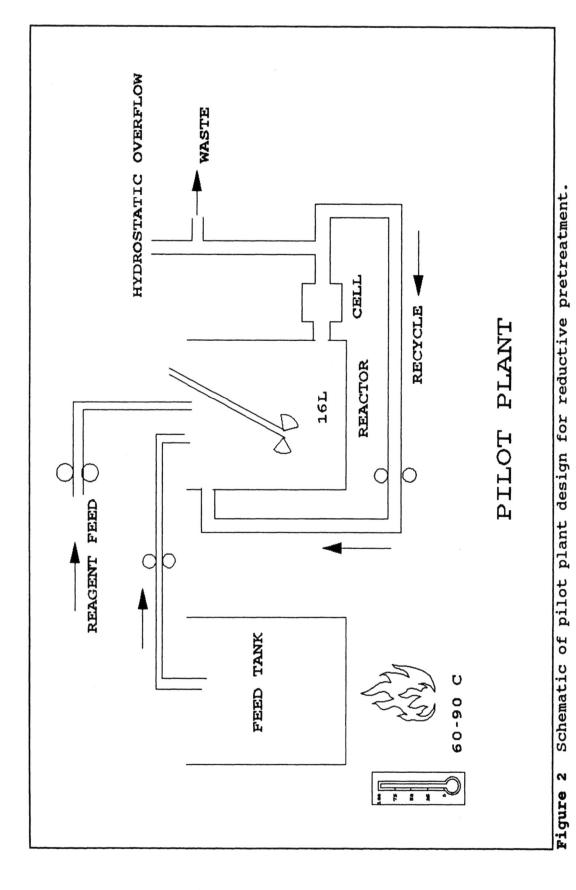

Figure 2 Schematic of pilot plant design for reductive pretreatment.

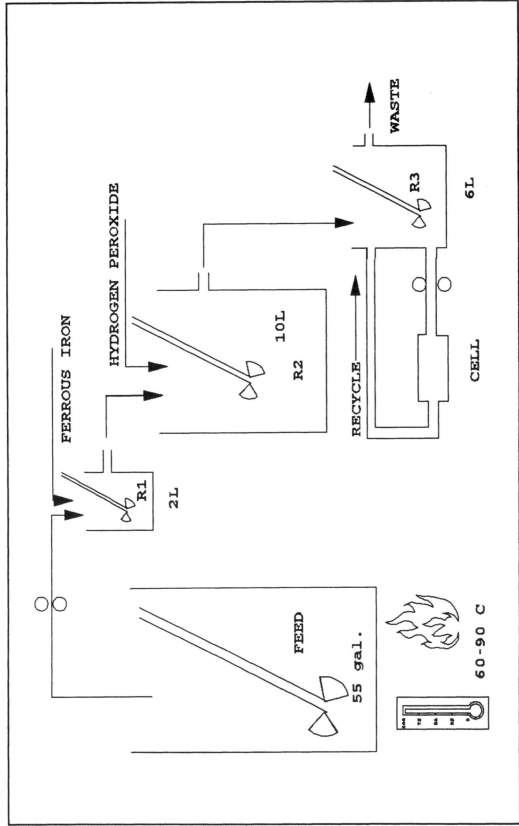

Figure 3 Schematic of pilot plant for oxidative pretreatment using Fenton's chemistry.

67

convert the response to a 0 to 5 V signal. The voltage is converted in a voltage to current converter which gives a 4 to 20 mA signal used to regulate the speed of a Masterflex pump. A flow diagram of the above sequence is shown in figure 4.

The entire system is mounted on a steel frame. The frame is fitted with casters and is completely portable. The circuitry was designed and built and the software was written by Riley Chan (Electrical Engineer for the Department).

The color measurement calculation used in the pilot plant is based on the ADMI Tristimulus Filter Method (Proposed) 2120E in APHA (1989). This method was chosen because of the restrictions placed on POTW'S by the EPA which stipulate ADMI units.

The ADMI method requires that light transmittance be measured at 3 different wavelengths, namely 590, 540, and 438 nm. The relation between the number generated by the pilot plant (0-4095) at each of the wavelengths was assumed to be linear when related to transmittance (100% - 0%) so the simple conversion

$TX = 100 - 0.024*X$

was used where

X = Transmittance for wavelength X from Pilot Plant (0 - 4095)

TX = % transmittance (100% - 0%).

The transmittances are used to calculate tristimulus values which relate the measured color to a position in color-space. Color-space uses a reference point; usually (0,0,0) in a three dimensional field in which the dimensions are darkness, red-green, and yellow-blue, respectively. Munsell values based on the tristimulus values are calculated using a curve-fit equation and the color value is calculated as the relative distance form the reference point in color space.

BIOLOGICAL REACTORS

Biological reactors were used to insure that the oxidative and reductive chemistry used in the pilot plant would not adversely affect any POTW (Publicly Owned Treatment Works) operating down-stream. Three continuous-flow biological rectors (CFBRs) were operated simultaneously at room temperature to perform aerobic biological treatment. A diagram showing the reactor set up is shown in figure 5. The biological reactors were made from 5 gallon buckets. A

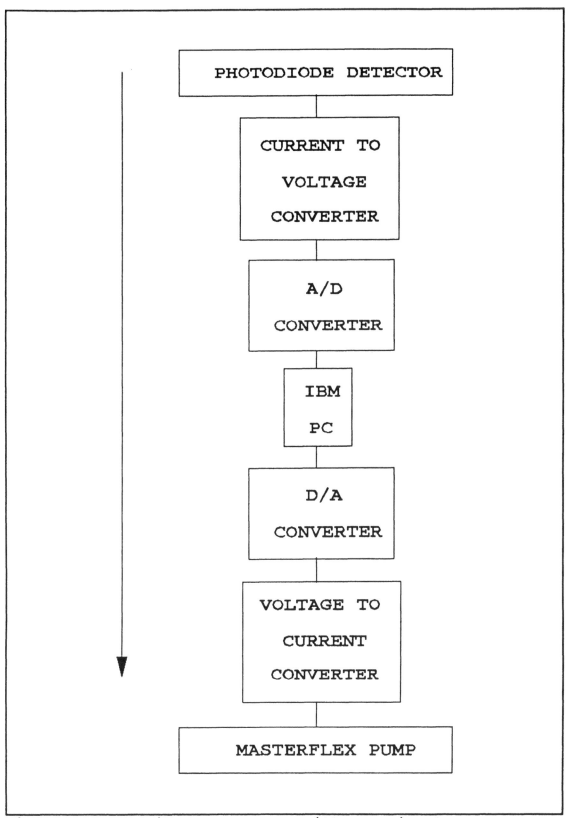

Figure 4 Flow diagram for transmittance signal.

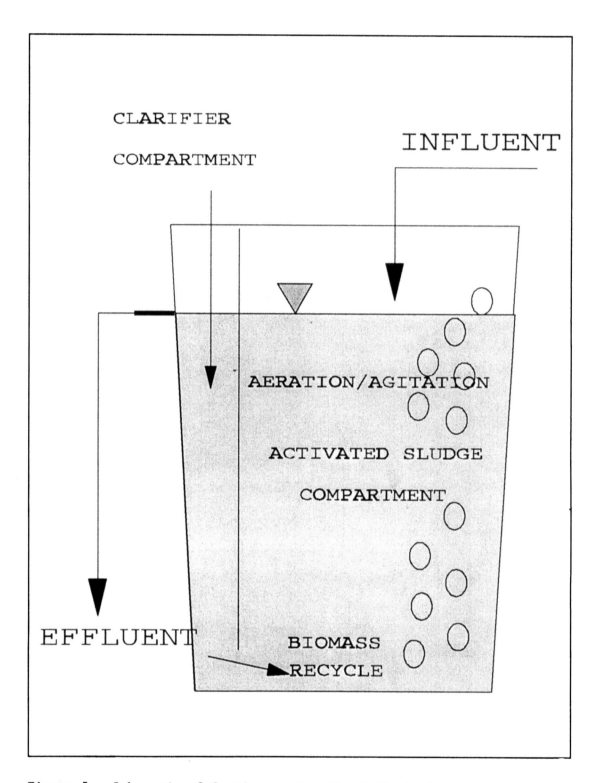

Figure 5. Schematic of Continuous Aerobic Biological Reactor.

baffle was placed inside the reactors which resulted in a smaller section that contained approximately 15% of the total reactor volume. This smaller section simulated a secondary clarifier and provided biomass recycle. The secondary clarify section was not aerated or agitated. The larger compartment in the reactors simulated the activated sludge process in a wastewater treatment plant. A drain was placed on the clarified side of the reactors to give an overflow volume of 12 L. A masterflex pump with three pump heads supplied feed to each of the three reactors at a constant rate of 2.1 mL/min. The feed to the reactors was maintained at 3 L per day which gave a 4 day hydraulic retention time. DOC measurements were used to monitor the reduction of organic carbon levels in the reactors.

The biological reactors were initially seeded with waste activated sludge that originated from the Martinsville, Virginia POTW. The microorganisms making up this sludge were assumed to be accustomed to exposure to the textile dyes used. To provide the necessary nutrients for the biomass, the dye waste was mixed with primary clarified waste form a local wastewater treatment plant. All three reactors were continuously aerated and agitated using lab air maintained at 20 psi.

MATERIALS AND METHODS

BIOREACTOR TEST PROCEDURES

Three aerobic biological reactors were maintained to observe the effects of the oxidized or reduced dye waste on biomass. The reactors were seeded with biomass which originated from the Martinsville, Va POTW. This biomass was chosen because of its "familiarity" with the type of dye waste being used in this experimentation.

The feed for the reactors was made in the ratio of 1:9:3 jet-dye waste to tap water to primary clarified sewage. This ratio provided for the feed mix to be 75%-10% jet-dye waste in water and 25% municipal waste. This feed mix was determined to best match actual conditions at the Martinsville, VA POTW (Lloyd, 1992). The primary clarified sewage was obtained from Pepper's Ferry, a local POTW. The Pepper's Ferry waste maintained an average BOD of 144 ppm and a suspended solids level of 100 ppm.

While waiting for the reactors to come to steady-state, all three reactors were fed from the same tank. When steady state was achieved, additional feed tanks were added and the feed formula was changed so that one reactor continued to treat the original feed formula and the other reactors had oxidized or reduced dye waste substituted for the jet-dye waste portion of the formula.

Samples were taken for analysis twice weekly. 50 ml

was removed from each clarified compartment, each activated sludge compartment and each feed tank. Each sample was analyzed for pH and color (ADMI). DOC was measured on feed samples and samples from the clarified compartment, and TSS solids was measured on feed samples and samples from the activated sludge compartment.

ON SITE PILOT PLANT OPERATIONS

The pilot plant was set up and operated on site at Tultex in Martinsville, Virginia. Three different fresh jet-dye waste concentrates were used in the pilot plant to demonstrate its effectiveness and to observe operation with several different color formulas. The dye formulas treated included Navy 106, an azo based Red formulation , and a phthalocyanine based formulation, Ming Jade.

The jet-dye waste was pumped into the pilot plant feed tank. There its pH was adjusted to 3 using 70 percent sulfuric acid, it was stirred and heated to 65°C. After calibration of the light sensing circuitry, operation of the pilot plant began. The feed and recycle flow rate were both set to 1.1 L/min. A 6000 ppm hydrogen peroxide solution was made using bulk 50 percent peroxide from Tultex. A corresponding Ferrous iron solution (300 ppm) was made by dissolving 29 g of $FeSO_4 \cdot 7H_2O$. The pilot plant was operated for several hours with each dye waste and data was collected for analysis of color and DOC removal and for cost of operation.

RESULTS AND DISCUSSION

REDUCTION OF SLACK WASHER WASTE

The first trials with the pilot plant on campus were run on Navy 106 slack washer waste. These trials were designed to test the ability of the control system to operate as designed and to get some initial reduction data.

The slack washer waste was heated to 60°C, using an electric emersion heater, to return it to the actual conditions found in the textile plant. A 6000 ppm solution of thiourea dioxide was prepared for use as the pretreatment reagent. The control circuitry was allowed to warm-up for 30 minutes and then the null and gain of the "spectrophotometer" were calibrated.

After calibration was completed and the dye waste had reached the prescribed temperature the feed/circulation pump was started. This began the dye waste flow into the reactor and the recycle stream through the light cell, both at 1 L/min. At the main menu of the control program, "P" was selected for PID control of the reactor. The set point was set to 400 ADMI units and the system was allowed to run

continuously. After approximately 15 minutes the reactor
volume reached its maximum and treated dye waste began to
leave the system through the hydrostatic overflow at the
same rate that fresh dye waste was being fed to the reactor
(1 L/min).

The results of this trial are shown in figure 6. This
graph shows the control variable (ADMI) vs run time. The
initial ADMI value of the dye waste according to the pilot
plant was 2980. The setpoint was reached 20 minutes after
the reactor was started. This is just longer than 1
residence time in the reactor. The control variable
oscillated about the setpoint with a decreasing amplitude,
showing that the controller was working and that the control
was underdamped. No biological data was collected for
reduced slack washer waste and all pilot plant work
following this run was done with concentrated jet-dye waste.

REDUCTION OF CONCENTRATED NAVY 106 JET-DYE WASTE

It was determined during the course of this research that it
would be advantageous to study the pretreatment of jet-dye
waste discharges because of the extremely high contribution
to overall color of the mill effluent relative to volume
added. This presented some problems in color measurement
that had to be addressed. The ADMI calculation
was designed to be effective in the range of 0 to 2000 ADMI.
Because the jet-dye waste being treated has an initial color
of from 55,000 to 75,000 ADMI and dilution was not
practical, alternative solutions had to be looked at.

One obvious solution was to decrease the path length
for the spectrophotometer. With a 70% decrease in the path
length, the response was easily adapted for in the control
software. The second solution was to abandon the ADMI
calculation and simply choose one of the three wavelengths
being measured and use it as the feed back control variable.
In the end, it was determined that a combination of the
narrower cell and the use of a single wavelength was best
for control of the Pilot Plant.

THIOUREA DIOXIDE REDUCTION OF CONCENTRATED NAVY 106 JET-DYE
WASTE

Figure 7 shows data collected for two residence times
of reductive pretreatment using thiourea dioxide at an
approximate concentration of 300 ppm. The graph shows the
three individual wavelengths represented in computer units
(not percent transmittance) plotted verses time. The
concentrated jet-dye waste was too dark for use of the ADMI
calculation, so this data was used to determine which
wavelength gave the best response for control purposesand
what value should be used for the setpoint for PID

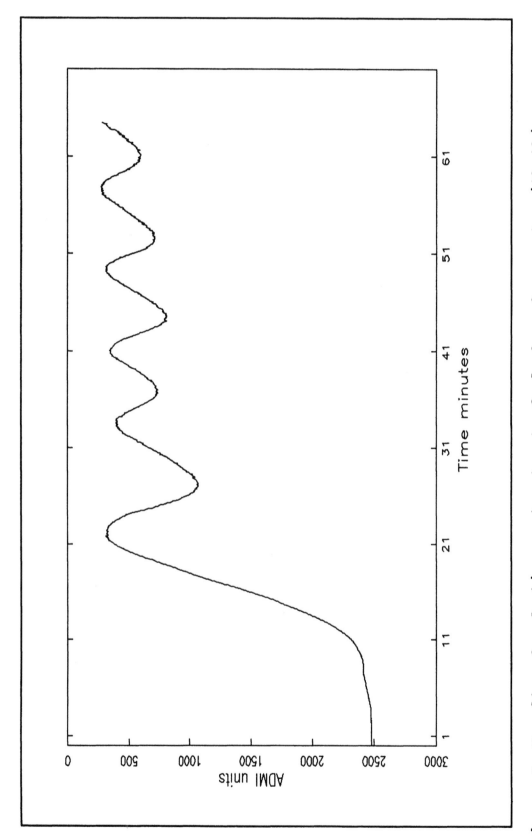

Figure 6 Results of reductive pretreatment of slack washer waste with thiourea dioxide in the Pilot Plant used to verify operation of control system. Setpoint at 400 ADMI.

74

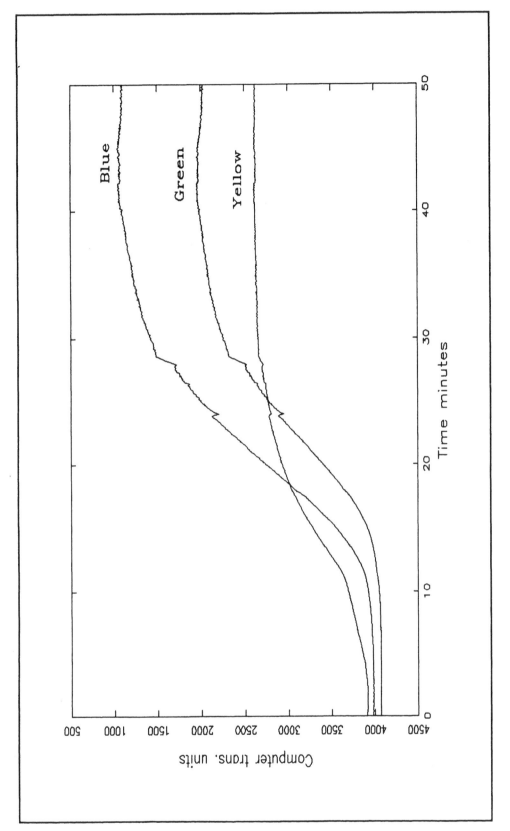

Figure 7 Graph showing the response of each wavelength vs time for reductive pretreatment of concentrated jet-dye waste using thiourea dioxide. From this graph it was determined that the blue wavelength would suit best for the control variable for this type of pretreatment.control. From this data it was determined that the

blue wavelength would be best for reactor control of reductive pretreatment because of the wide range of its response. This was probably due to the blue color of the particular dye formula used in this research; one of the other wavelengths may be

advantageous for red or yellow shades. A blue transmittance of 1200 was determined from figure 7 to be the optimum set point setting for the controller.

The color of the waste resulting from the thiourea dioxide pretreatment was measured immediately without pH adjustment or filtration in the laboratory spectrophotometer and found to be 3900 ADMI. This value corresponds to the lowest level attainable in bench scale batch trials. The resulting stream was a rich yellow color (which contributed to the 3900 ADMI value) that resulted from the reduction products (amines) of the concentrated jet-dye waste. The sample was aerated (oxygenated) for 3 minutes and the result was a change in color from yellow to deep purple. The sample was then diluted with distilled water, filtered, and the ADMI value was read on the laboratory spectrophotometer. The ADMI value went up to 8987 which clearly demonstrates color return. The waste resulting from the reduction trial was stored in a refrigerator and used for feed material to the biological reactors.

SODIUM HYDROSULFITE REDUCTION OF CONCENTRATED NAVY 106 JET-DYE WASTE

Figure 8 shows data collected for two residence times of reductive pretreatment using sodium hydrosulfite at an approximate concentration of 600 ppm. The graph shows the three individual wavelengths plotted verses time. Sodium hydrosulfite was not effective for color removal in the pilot plant and was not used further.

OXIDATION

Lab Tests of Fenton's Oxidation

Figure 9 shows the response of the individual wavelengths verses time for oxidative pretreatment of Navy 106. The data shows that the green or yellow wavelength is best for reactor control.

The color of the waste resulting from the Fenton's reagent pretreatment was measured in the laboratory spectrophotometer and found to be 577 ADMI. This is a reduction of 98.8% relative to the concentrated jet-dye waste. There was no return of color observed in any oxidized samples. The resulting waste stream from the oxidative pretreatment was stored in a refrigerator and used as feed material for biological treatment.

76

Biological Testing Results

Table 1 shows the steady state conditions of the biological reactors prior to the introduction of pretreated waste. The reactor feed contained 75 percent concentrated navy 106 jet-dye waste diluted 1:10 with tap water and 25 percent primary clarified waste from a local POTW. The data represent a 12 day average which includes 4 hydraulic retention times. The data indicate that all three reactors were successfully degrading waste as is evidenced by the roughly 55 percent decrease in organic carbon levels. Color removal, as shown in previous research (Powell, et al., 1992), was not significant in the aerobic reactors. The reactors were allowed to operate for 2 months before steady state conditions were reached.

Figure 10 represents the results of biological action on the jet-dye waste after pretreatment in the pilot plant. All three biological reactors are represented with reactor 1 being the control, reactor 2 treating oxidized dye waste treated with Fenton's reagent, and reactor 3 treating reduced dye waste treated with thiourea dioxide. All reactor conditions were kept constant except that the feed to reactor 2 contained 75 percent oxidized dye waste diluted 1:10 with tap water, and the feed to reactor 3 contained 75 percent reduced jet-dye waste diluted 1:10 with tap water. The results show that oxidative pre-treatment removed 98.8% of the color in the pilot plant and had an over-all color removal after aerobic digestion in bio-reactor 2 of 96.8%.

The concentrated jet-dye waste that was reductively pre-treated had a minimal color value of 3900 ADMI which was a 92.2% reduction in color from the concentrated dye waste. The over-all color removal after aerobic digestion in bio-reactor 3 was 76.6%. The pilot plant effluent stream from reductive pre-treatment was a deep yellow color and upon dilution, prior to biological treatment, a significant amount of blue color returned and was not removed significantly in biological treatment resulting in the decrease in the percent color removal. Figure 11 shows the results of DOC removal after oxidative and reductive pretreatment and is intended as an indication of biological activity.

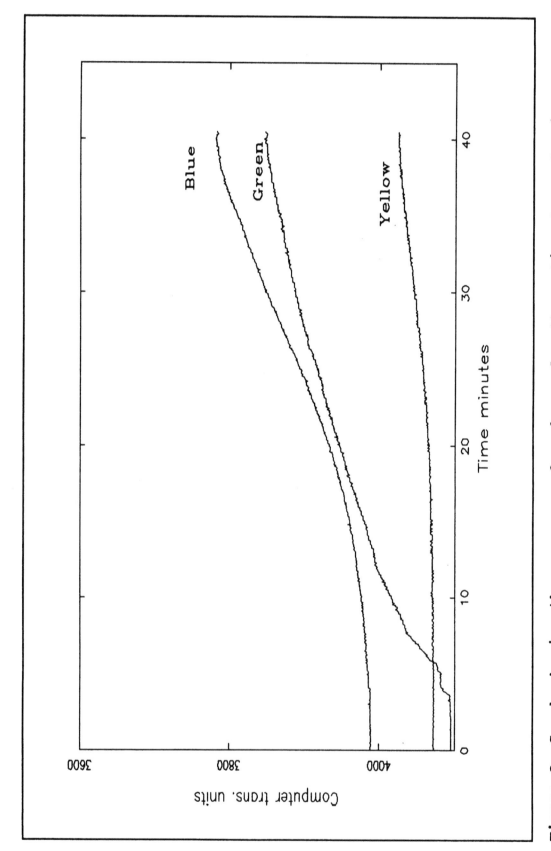

Figure 8 Graph showing the response of each wavelength vs time for reductive pretreatment using sodium hydrosulfite.

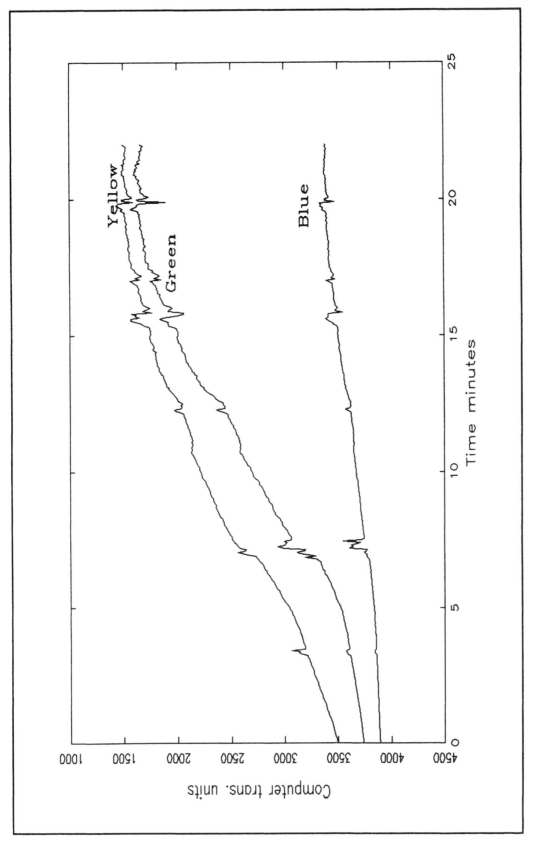

Figure 9 Graph showing response to oxidative treatment.

79

Table 1 Steady State Conditions of Biological Reactors. Data Represent 12 Day (4 Hydraulic Retention Times) Average. Feed Contains 75% Untreated Jet-dye Waste at 10% Concentration and 25% Municipal Waste and Was Fed in Equal Amounts Simultaneously to all Reactors.

	ADMI	pH	TC	IC	DOC	TSS
PURE JET-DYE WASTE	50,133	12.16	777.3	461.4	315.9	188
FEED	3650	7.5	113	69.97	43.03	50.2
BIO-REACTOR 1	3592	8.1	50.47	31.27	19.2	276.3
BIO-REACTOR 2	3628	8.1	52.69	33.03	19.66	288.7
BIO-REACTOR 3	3613	8.3	51.88	32.5	19.3	290.4

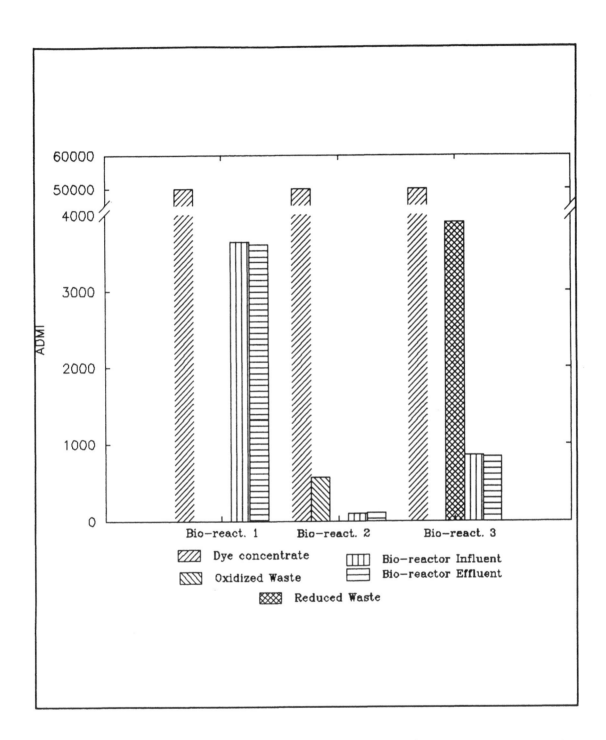

Figure 10 Effects of Oxidative and Reductive Pre-treatment on the Removal of Color From Concentrated Jet-dye Waste in Biological Reactors. Reactor 1 is the Control Reactor, Reactor 2 Represents the Aerobic Treatment of Oxidized Jet-dye Waste, and Reactor 3 Represents the Aerobic Treatment of Reduced Jet-dye Waste.

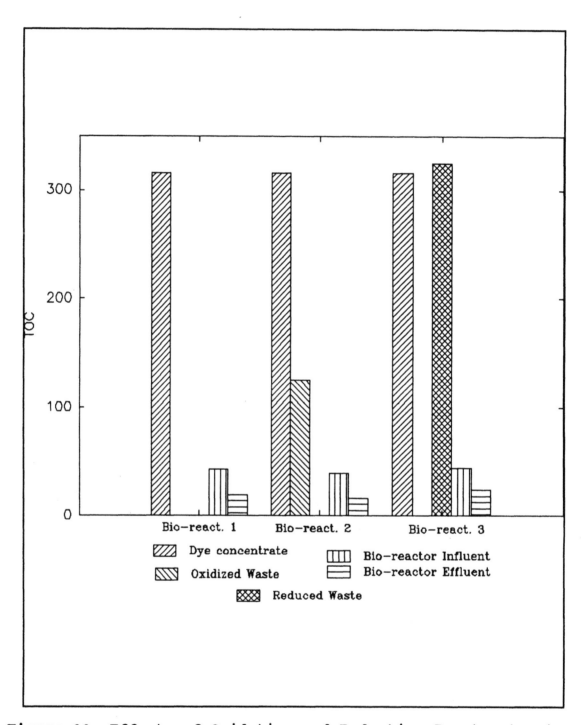

Figure 11 Effects of Oxidative and Reductive Pre-treatment on the Removal of DOC From Concentrated Jet-dye Waste in Biological Reactors. Reactor 1 is the Control Reactor, Reactor 2 Represents the Aerobic Treatment of Oxidized Jet-dye Waste, and Reactor 3 Represents the Aerobic Treatment of Reduced Jet-dye Waste.

On Site Pilot Plant Operation

The following results were obtained from on site Pilot Plant operations at Tultex in Martinsville, Va. Three different jet-dye concentrate formulas were tested in the pilot plant for color removal. They included Navy 106 jet-dye concentrate, an azo-based Red jet-dye concentrate, and a copper phthalocyanine based Ming Jade.

The Pilot Plant was transported to the textile mill and set up on sit for continuous treatment of fresh jet-dye waste. The waste was pumped from the jet into a 55 gallon tank after the dye cycle was complete. Low pressure steam was used to keep the dye at the exit temperature of 60 °C. The pH of the dye waste was adjusted to 3.0 using 70 percent sulfuric acid and pilot plant operation was begun.

The initial ADMI values as read on the laboratory spectrophotometer were 48,432, 59,757, and 114,990 for the navy, red, and jade, respectively.

ON SITE TREATMENT OF NAVY 106

Figure 12 shows the three wavelengths vs time resulting from the treatment of Navy 106 jet-dye concentrate. The green wavelength was used as the controlled variable and the setpoint was set at 2800. Although this was not the lowest level attainable, it was chosen to allow the pilot plant to reach a steady state sooner so that data for reagent consumption could be obtained. Although the waste stream from the Pilot Plant maintained a darkish gray-brown color, a sample taken from that stream which was quenched with sodium hydrosulfite and filtered using a syringe and a .22 μm showed that the color was effectively removed. The ADMI of the filtered sample was measured on the laboratory spectrophotometer and found to be 822 ADMI units which represents a 98% reduction in color.

Table 2 summarizes the cost involved in treating the Navy 106 jet-dye concentrate. Data for reagent consumption was obtained by recording the volume of reagent used during a fixed time period. Data on sulfuric acid consumption was based on the amount of 70% sulfuric acid required to reduce the pH of the jet-dye concentrate to 3 in the 55 gallon feed tank. No data on the cost of heating and maintaining the feed tank at 60°C using the mill's steam was obtained.

On Site Treatment of Ming Jade

Figure 13 shows the results of the treatment of the Ming Jade jet-dye concentrate waste. The run is shorter because not as much jet-dye waste was obtained for treatment.

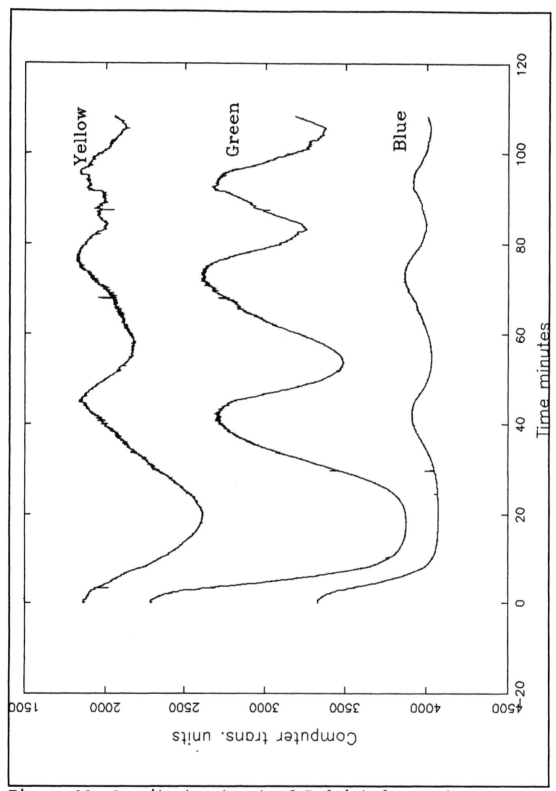

Figure 12   On site treatment of Red jet-dye waste

Table 2 Cost summary for treatment of Navy 106 jet-dye waste concentrate at Tultex. Prices are from Chemical Management Review June 7, 1993. The cost for H_2O_2 is from Tultex, Martinsville, Virginia.

Reagent	Consumption	Cost	Total
H_2SO4 (70%)	39.5g/gal	$0.0375/lb	$3.25/1000gal
H_2O_2 (50%)	0.0084 lbs/gal	$0.4/lb	$3.37/1000gal
$FeSO_4*7H_2O$	.447g/gal	$0.005/lb	$0.0045/1000gal
			$6.63/1000gal

ON SITE TREATMENT OF RED

Figure 14 shows the results of the on site treatment of the red jet-dye concentrate. Once again the green wavelength was used as the control variable and the setpoint was set at 3000. A sample of the pilot plant waste stream was recovered and quenched with sodium hydrosulfite. The sample was then syringed through a 0.22 μm filter and the color was measured on the laboratory spectrophotometer. The color was found to be 5690 ADMI units which represents a 90% reduction in color. Table 3 shows the cost data involved in treating the Red jet-dye concentrate waste.

Table 3 Cost summary for treatment of an azo based Red jet-dye waste concentrate at Tultex. Prices are from Chemical Management Review June 7, 1993. The cost for H_2O_2 is from Tultex, Martinsville, Virginia.

Reagent	Consumption	Cost	Total
H_2SO4 (70%)	21.27g/gal	$0.0375/lb	$1.76/1000gal
H_2O_2 (50%)	0.0132 lb/gal	$0.4/lb	$5.28/1000gal
$FeSO_4$	0.698 g/gal	$0.005/g	$0.007/1000gal
			$7.05/1000gal

It appeared at first that the pilot plant was not effectively treating the Ming Jade waste because there was very little color change between the feed line to the pilot plant and the overflow waste stream. However, when a sample of the waste stream was quenched and filtered it was obvious that most of the apparent color could be attributed to solids. The color of the filtered sample was measured on

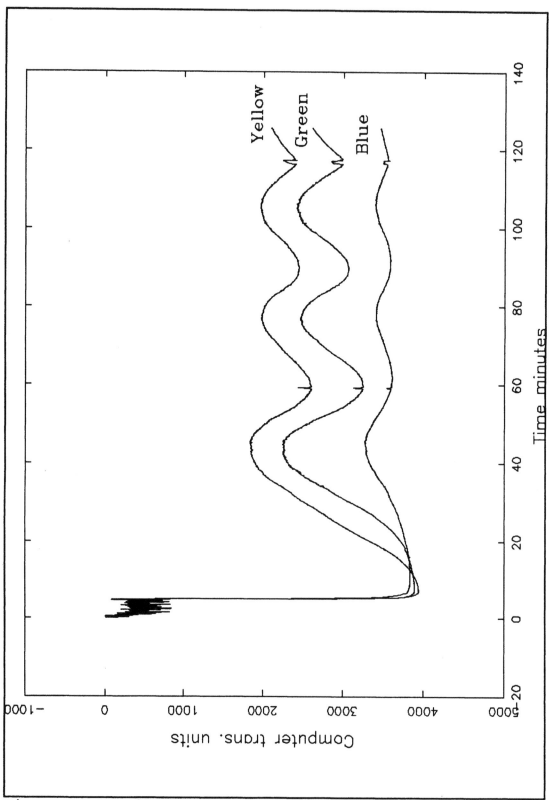

Figure 13 On site treatment of Navy 106 jet-dye

the laboratory spectrophotometer and found to be 1320 ADMI units which represents a 99% reduction in color. Also, the copper concentration was measured for the treated filtered sample and for the untreated sample in a flame atomic absorption spectrophotometer. The untreated sample had a value of 19.22 ppm Cu while the treated and filtered sample had a concentration of 4.5 ppm, a 75% reduction in copper concentration.

Because the solids concentration was so high upon treatment in the pilot plant, the value for cost of operation shown in table 4 is not representative of the actual cost as the reagent feed ran continuously.

Table 4 Cost summary for treatment of a copper phthalocyanine containing Ming Jade jet-dye waste concentrate at Tultex. Prices are from Chemical Management Review June 7, 1993. The cost for H_2O_2 is from Tultex, Martinsville, Virginia.

Reagent	Consumption	Cost	Total
H_2SO4 (70%)	26.96g/gal	$0.0375/lb	$2.23/1000gal
H_2O_2 (50%)	0.0102lb/gal	$0.4/lb	$5.28/1000gal
$FeSO_4$	0.54g/gal	$.005/lb	$0.006/1000gal
			$7.52/1000gal

Table 5 shows data on DOC and TSS for the three jet dye concentrates before and after treatment.

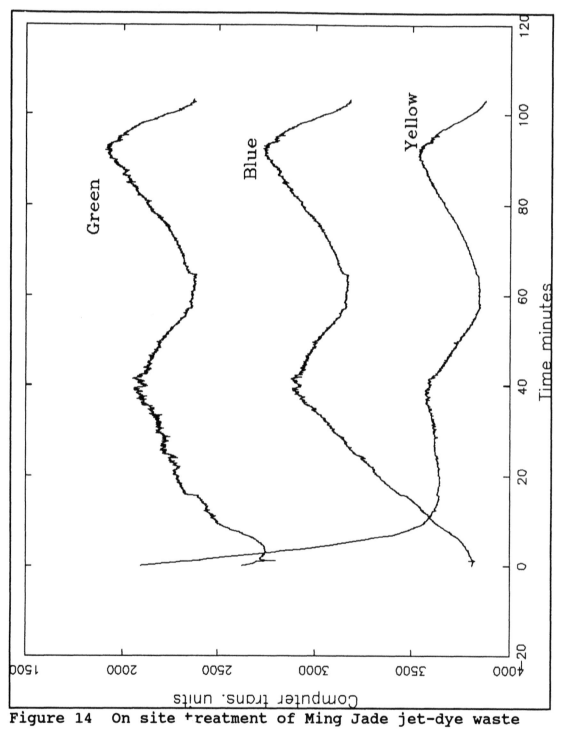

Figure 14   On site treatment of Ming Jade jet-dye waste

88

Table 5 Total Organic Carbon and Total Suspended Solids
data for dyes treated on site at Tultex, Martinsville,
Virginia.

Sample	DOC ppm	TSS g/l
Navy before treatment	515.2	0.988
Navy after treatment	500.0	1.5
Red before treatment	452.2	0.732
Red after treatment	368	1.766
Jade before treatment	432.2	0.575
Jade after treatment	350.4	1.505 g/L

CONCLUSIONS

1.   Oxidation in the form of Fenton's chemistry is more effective at color removal on a continuous basis than reducing chemicals (thiourea dioxide or sodium hydrosulfite) in the pilot plant. Fenton's reagent achieved color removals of 98 and 90 percent for continuous treatment of jet-dye waste concentrates including Navy 106 and an azo based Red color, respectively, and 99 percent color removal for a phthalocyanine, Ming Jade. Also, color return upon aeration was noted in samples treated with reduction chemistry so that color removal after aeration for reduction chemistry was 76.6%.

2.   DOC reduction of 50 to 55 percent in continuously operated biological reactors indicated that the reaction products of oxidative and reductive chemistry of Navy 106 jet-dye concentrate did not increase toxicity to biological activity.

3.   Color control of oxidative and reductive pretreatment of reactive dye waste water streams is practical. ADMI color value control is possible on those waste water streams with an initial color value below 3000 ADMI. Above 3000 ADMI, transmittance at a single wave length is a viable method of control with the best wavelength being dependent on the color of the waste being treated.

4.   Oxidative treatment decreased soluble copper concentration in a copper containing waste stream by 76 percent from 19.2 ppm to 4.5 ppm, thereby increasing the total suspended solids in that waste by 62 percent.

5.   Color removal of 98 percent was attained for Navy 106 jet-dye waste concentrate using Fenton's chemistry at an approximate chemical cost of $6.60 per 1000 gallons.

LITERATURE CITED

American Public Health Association; American Water Works Association; Water Pollution Control Federation, 1989, "Standard Methods for the Examination of Water and Wastewater," 17th ed., American Public Health Association, Washington, DC.

Balko, J., 1992, School of Chemical Engineering, Virginia Polytechnic Institute and State University, Laboratory Notebook and Personal Communication.

Bell, C.F., A.M. Dietrich, R.D. Voyksner, and D.L. Michelsen, 1992, "Monitoring for Sulfonated Azo Textile Dyes and Their Degradation Products," Abstract Presented Before the Division of Environmental Chemistry American Chemical Society, Washington, D.C., August.

Linneman, R.C. and T.H. Flippin, 1991, "Hydrogen Peroxide Treatment of Inhibitory Wastestream: Bench-Scale Treatability Testing to Full-Scale Implementation: A Case History," Proceedings to the First International Symposium on Chemical Oxidation, Nashville, TN. Technomic.

Loyd, Kemper, 1992, "Anaerobic/Aerobic Degradation of a Textile Dye Wastewater," Virginia Polytechnic Institute, Blacksburg, VA, March.

Mann, J., and R.M. Woodby, 1991, "Pilot Plant for Chemical Treatment of Navy 106 Dye Wash Water," Report to Dye Project Review Committee, April.

McCurdy, M.W., G.D. Boardman, D.L. Michelsen, and R.M. Woodby, 1991, "Chemical Reduction and Oxidation Combined with Biodegradation for the Treatment of a Textile Dye Wastewater," 46th Annual Purdue Ind. Waste Conf., May.

Meyer, U., 1981, "Biodegradation of Synthetic Organic Colorants," In: "Microbial Degradation of Xenobiotics and Recalcitrant Compounds," T. Leisinger, R. Hutter, A.M. Cook, and J. Nuesch, eds., Academic Press, Inc., New York, NY.

Michelsen, D.L., L.L. Fulk, R.M. Wookby, and G.D. Boardman, 1991, "Adsorptive and Chemical Pretreatment of Reactive Dye Discharges," ACS I&EC: Symposium on Emerging Tech. for Haz. Waste Mgmt., Atlanta, GA, October.

Pagga, U., and D. Brown, 1986, " The Degradation of Dyestuffs: Part III - Behavior of Dyestuffs in Aerobic Biodegradation Tests," Chemosphere, 15(4):479-491.

Powell, W.W., D.L. Michelsen, G.D. Boardman, A.M. Deitrich, and R.M. Woodby, 1992, "Removal of Color and TOC From Segregated Dye Discharges Using Ozone and Fenton's Reagent," 2nd International Symposium Chemical Oxidation Technology, Vanderbilt University, Nashville, TN, February.

Tincher, W.C., 1991, School of Textile and Fiber Engineering, Georgia Tech, Personal Communication.

Weber, E.J., and N.L. Wolfe, 1987, "Kinetic Studies of Reduction of Aromatic Azo Compounds in Anaerobic Sediment/Water Systems," Environmental Toxicology and Chemistry, 6(4):911-920, June.

L. LI
D. H. CHEN
K. Y. LI
J. A. COLAPRET

Nitric Oxide Removal with Ozone in a Bubbling Reactor

ABSTRACT

NO_x emission control is needed for all stationary sources to comply with the Clean Air Act Amendment of 1990 and state regulations. The main obstacle of NO_x removal using wet scrubbing is, however, the low solubility of nitric oxide (NO), the main constituent of NO_x. In this study, the experimental efforts were directed toward enhancing NO_x absorption in scrubbing liquors by the oxidation of NO with O_3 in the liquid phase.

Experiments were conducted by passing O_3/O_2 through a bubbling reactor to remove 200-2000 ppm NO. More than 99% NO removal efficiency has been observed in liquid phase oxidation by O_3/O_2 at pH=13. The effects of UV, pH, temperature, concentrations (NO, O_2, O_3) were studied. Whereas NO removal efficiencies by O_2 appear to be unchanged with different pH values, NO removal by O_3 increases as the pH values vary from 1 to 13 using various reagents (HCl, HNO_3, buffer solution, $NaHCO_3$, Na_2CO_3, and NaOH). The effect of scrubbing liquor temperature on NO removal from 22 °C to 70 °C is significant in ozonation. A lower operating temperature favors the removal of NO. Ozone is found to be more effective at lower NO concentrations, where a better utilization of O_3 was observed. The liquor in the bubbling reactor after NO absorption was analyzed by the Hach DR/3000 spectrophotometric method. Mole concentration of NO_2^- increases linearly with time. About two thirds of the removed NO were converted to $[NO_2^-]$ and one third converted to $[NO_3^-]$.

The rate of liquid phase NO oxidation was found to be second order in NO and first order in O_2 which is similar to the gas phase oxidation. The apparent rate constant k_{LO2} for liquid phase oxidation of NO was found to be 1.4 x 10^{-9} ppm^{-2} sec^{-1}, which is about 60 times higher than the gas phase rate constant k_{GO2}, 2.34 x 10^{-11} ppm^{-2} sec^{-1}. The rates of NO ozonation in both the gas and liquid phases follow a second-order reaction. The apparent rate constant k_{LO3} was found to be 0.6 ppm^{-1} sec^{-1} with the stoichiometric ratio of NO to O_3 being 1:6 in the liquid phase reaction. The value of k_{GO3} was found to be 0.005 ppm^{-1} sec^{-1} with the stoichiometric ratio of NO to O_3 being 1:3 in the gas phase reaction. The larger apparent rate constants K_{LO3} and K_{LO2} indicate that NO, the main insoluble part of NO_x, can be removed with a faster kinetics or smaller reactor volume. The high stoichiometric O_3 to NO ratio means less O_3 is required for NO_x removal in liquid phase operations. The presence of O_2 is also beneficial in the liquid phase removal of NO.

Environmental Chemistry Laboratory, Lamar University, Beaumont, TX, 77710

INTRODUCTION

The air pollution caused by nitrogen oxides (NO_x) is a major concern in the U. S. and throughout the industrialized nations. NO_x is not only responsible for acid rain but also for urban area smog through photochemical reactions with hydrocarbons. Exposure of NO_x also affects human health in term of susceptibility to disease [7,10]. In recognition of these problems, the federal and state regulations for NO_x control have become more and more stringent. Under the Clean Air Act Amendment of 1990, NO_x emission needs to be reduced to 2 million tons below the 1980 level and the maximum achievable control technology (MACT) will have to be implemented on all future sources by the year 2000 [6,8]. The state of Texas now requires sources that emit in excess 100 tons per year of NO_x must inventory their emissions and allows only 9 ppm emission in flue gases in some operating permits [34].

NO_x normally can refer to all of the oxides of nitrogen. In air pollution control, however, NO_x generally refers only to nitric oxide (NO) and nitrogen dioxide (NO_2) because they are the major species in NO_x emissions. In the United State, about 60 % of nitrogen oxides are found in flue gas (or tail gas) from stationary sources such as industrial furnaces, process heaters, utility boilers, incinerators, and metallurgical/ nitric acid plants. NO_x gases are generally present in the flue gases in the range of 400-2000 ppm. In these emissions, nitric oxide, the principal form of NO_x in flue gas (accounting for more than 90 % of all emitted NO_x), is comparatively stable and unreactive. As a result, the technology for NO_x control is not as advanced as for SO_x and is still a field of active research.

For control of NO_x emissions, two types of processes are being developed: combustion modifications and flue-gas (tail-gas) treatment processes. Among combustion modifications, gas recirculation, steam injection, gas reburning, and low NO_x burners are the most common [8,11,26]. In flue gas treatment, there are dry and wet processes. Dry processes include selective catalytic reduction (SCR) of NO_x to N_2 with ammonia, non-selective catalytic reduction (NSCR) of NO_x to N_2 with H_2, CO, and hydrocarbons, selective non-catalytic reduction (SNCR) with ammonia, and adsorption. SCR is most often recommended to achieve high level NO_x control but is generally considered as expensive [11,18,20,25,35,36]. Wet processes include absorption with liquid phase oxidation, absorption with liquid phase reduction, and gas phase oxidation followed by absorption [3,5,9,18,23]. In wet processes, the most promising technology is to oxidize nitric oxide to more soluble and more reactive nitrogen dioxide. Since wet scrubbing systems have been installed in many plants for desulfurization, a denitrification process can be coupled with the existing desulfurization process to achieve combined removal of NO_x and SO_2 [20,24,33]. Many recent research efforts in the flue gas cleanup technology have been devoted to enhance NO_x absorption in scrubbing liquor, including the oxidation of NO with strong oxidants such as O_3, ClO_2, and H_2O_2 [1,2,17].

Ozone is a powerful oxidizing agent and has been widely used in water and waste water treatment with a great deal of success [21,22,27,30]. Although much work has been done in Japan with ozonation in gas phase for flue gas cleanup, little work on NO_x ozonation in liquid phase has been reported. The main obstacle for wet scrubbing of NO_x is the low solubility of NO. In order to effectively absorb NO at normal flue gas concentrations, chemical enhancement is required [20]. One technique for chemical enhancement involves the gas phase or liquid phase oxidation of NO to NO_2 followed by absorption of NO_2 in aqueous scrubbing solution. Another technique is absorption of NO_x with liquid phase reduction by using reactive and complex forming solvents.

In this study, the experimental efforts were directed toward exploring the potential of liquid phase O_3/O_2 reaction with NO. Experiments were carried out with a liquid scrubbing system in which the liquid is in a semi-batch mode while the gas is running in a continuous mode. NaOH as well as other reagents was added to the scrubbing liquor. A chemiluminescent NO_x analyzer and a sample drying system were installed to measure the outlet concentration of NO or NO_x on line. The gas phase reactions were also conducted to compare with liquid phase reactions. The effects of UV, pH, temperature, and concentrations (NO, O_3, and O_2) on NO removal were examined. The effectiveness of liquid phase oxidation by O_2/O_3 under various operating conditions was evaluated.

EXPERIMENTAL

The experiments were conducted using a laboratory scrubbing apparatus as shown in Figure 1. The liquid is in a semi-batch mode while the gas is running at a continuous mode. The gas mixture then goes through another gas washing bottle containing a potassium iodide (KI) solution and finally a sampling system. The NO and NO_x concentrations in the outlet gas were measured by a NO_x analyzer on line. In a typical experiment, NaOH scrubbing liquid was filled in a standard gas washing bottle (Fisher's, 500 ml). The NO gas from a cylinder certified as 1.00% NO in N_2 was first diluted in a gas distributor with nitrogen. Another stream of pure oxygen, passing though ozonator (Weslbach Model L-5620) was then mixed with the diluted NO in the same gas distributor. The relationship among oxygen flow rate, power setting, and O_3 concentration was calibrated prior to the experiments. The gas mixture was bubbled through the fretted dispersor at the bottom of the bubbling reactor at a flow rate of 0.6-1.0 L min^{-1} and from there into a second gas washing bottle containing 0.5 L of a standard KI solution. Upon

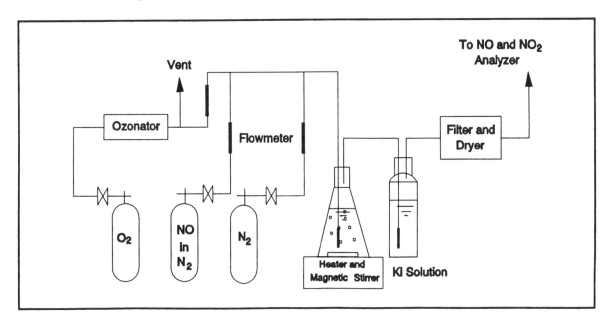

Figure 1 A Liquid Scrubbing System to Remove Nitric Oxide

bubbling through the KI solution, the left ozone trace in the gas mixture was absorbed in the KI aqueous solution for removal and analysis of O_3. The treated gas stream finally passed through a sampling system (Perma Pure, Model PD-625-24-SS) with proper temperature control which can prevent a dense moisture formed downstream. The NO and NO_x concentrations in the outlet gas were measured by a Thermo Environmental Model 10S chemiluminescent NO-NO_x analyzer. The instrument has to, prior to use, be calibrated with a gas cylinder containing a known sample of nitric oxide in nitrogen. The NO_2^- and NO_3^- concentrations in the spent NaOH solution were determined by a spectrophotometric method (Hach DR/3000 Procedure).

In order to compare the oxidation of NO with O_3 and O_2 in the gas and liquid phases, a small empty bottle (30 ml) was also used in lieu of the bubbling reactor. The bottle volume was determined by calculating the gas flow rate and the rise velocity of a bubble relative to liquid (0.3 m/s) so that the gas phase residence time can be kept constant. For some experiments aimed at studying the effect of scrubbing solution temperature, a flat-bottom bubbling reactor with a magnetic stirrer was placed on a heater. Tygon or Teflon plastic tubing was used throughout the system to avoid reactions with ozone in the gas stream.

Several rotameters were calibrated by a wet test gas meter (Catalog No. 03125, Precision Scientific) using pure nitrogen or oxygen at room temperatures. One flowmeter was calibrated with soap flowmeter using oxygen. Ozone concentrations at the inlet and outlet of the reactor were analyzed by the potassium iodide (KI) method. A known quantity of potassium arsenite solution was put in the KI absorptive solution with starch as an indicator. The amount of potassium arsenite left in the solution after ozone absorption was then determined by titration [31].

Various chemicals (NaOH, Na_2CO_3, $NaHCO_3$, buffer solution, HCl and HNO_3) were used to prepare the scrubbing solutions at room temperatures to study the effect of pH values on NO_x removal. A Fisher Accumet Model 950 pH Meter (Catalog No. 13-636-950 & 951) was used in this study to determine the pH value of solution before and after the reaction.

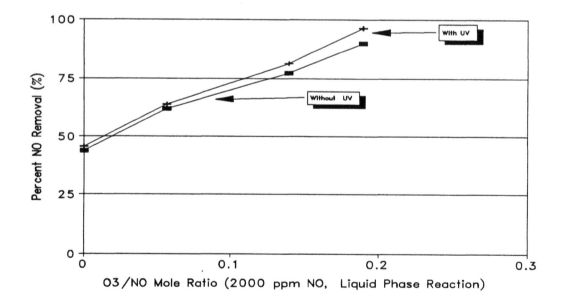

Figure 2 Effect of UV Radiation on NO Removal

RESULTS AND DISCUSSION

EFFECT OF OPERATING CONDITIONS

The removal efficiency of NO by UV/ozone in the liquid phase was studied. The gas phase reactions were also conducted to compare with liquid phase reactions. Dependance of NO removal efficiency on various operating conditions was also studied and the results are summarized below.

Effect of UV   It was found that the effect of UV radiation was insignificant, Figure 2. The reason may be the washing bottles used were not made of quartz, or ozone is quickly decomposed at alkaline solution, or even the OH radicals recombine quickly at high pH.

Effect of Oxygen Concentration   Figure 3 shows that increasing the concentration of oxygen improves the removal of NO. The slopes of these three curves increase slightly as O_3/NO ratio increases. This phenomenon indicates possible synergistic effect of O_2 and O_3 in the liquid phase NO removal.

Effect of pH   The influence of pH on the effectiveness for NO removal has been examined for the pH range of 1.0 -13.0 and the results are illustrated in Figure 4. Whereas NO removal efficiencies by O_2 appear to be the same in different pH value, the effect of pH on NO removal is very pronounced in ozonation as the pH values very from 1 to 13.   This interesting finding can be explained by O_3 decomposition in aqueous solution. The principal oxidant at low pH values is the molecular ozone, whereas at high pH, ozone quickly decomposes to free radicals and atomic oxygen. These species are very active oxidants.   In addition, the resulting NO_2 can be easily scrubbed in more alkaline solutions.

Effect of NaOH Concentration   absorbents with different NaOH concentrations (same pH) have been used in experiments to study the effect on NO removal. The results show that NO removal efficiency is essentially the same for NaOH concentrations in the range of 1-10 wt%.

Effect of Scrubbing Solution Temperature   Figure 5 shows the effect of NaOH solution temperature (22 - 70 °C) on the removal of NO. Higher O_3/NO ratios are needed for NO removal at higher temperatures than at room temperatures.   This is consistent with the rationale that ozonation in liquid phase comprises the transfer of gas phase ozone to the aqueous phase and the reaction of the dissolved ozone (or active radicals produced by decomposition of ozone) with NO. The transfer of ozone can be the controlling step if the ozonation reaction is very fast [13,21].   Increasing solution temperature causes a reduction in solubility of NO_x and O_3 which significantly effects the transfer of O_3 and NO_x to the liquid phase.

Effect of Inlet NO Concentration   Removal rate of NO at different NO concentration is shown in Figure 6.   The part of NO removal by O_2 is predominately controlled by the concentration of NO due to the third order reaction of NO with O_2 in liquid phase. Bubbling O_2/air through scrubbing liquid may have more beneficial effect on NO removal at higher concentrations of NO.   On the other hand, O_3 is found to be more effective at lower concentrations of NO, where better utilization of O_3 can be achieved.

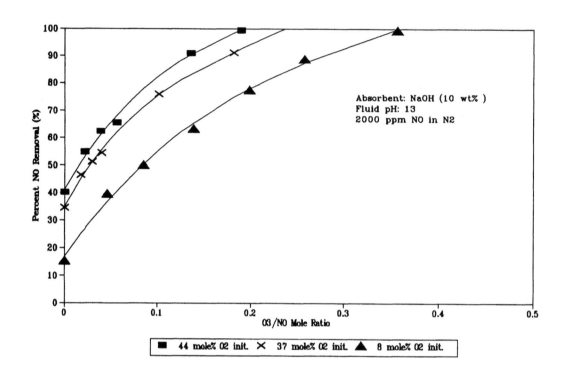

Figure 3 The Effect of Concentration of Oxygen

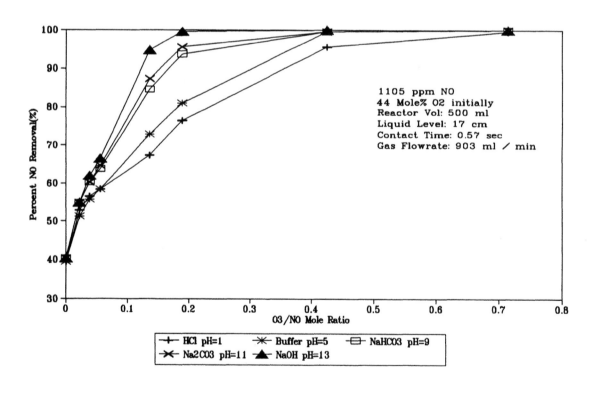

Figure 4 The Effect of pH

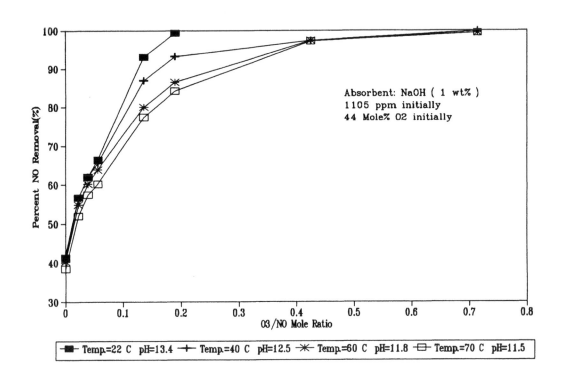

Figure 5 The Effect of Temperature of Scrubbing Solution

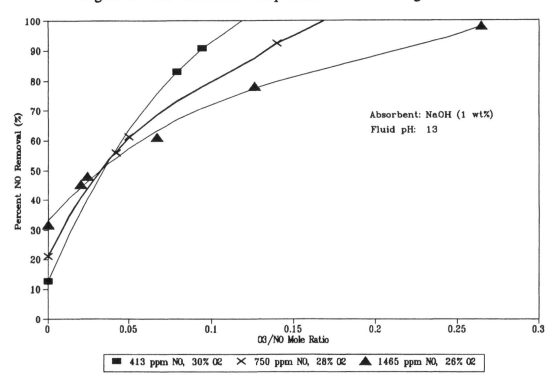

Figure 6 The Effect of NO Concentration

99

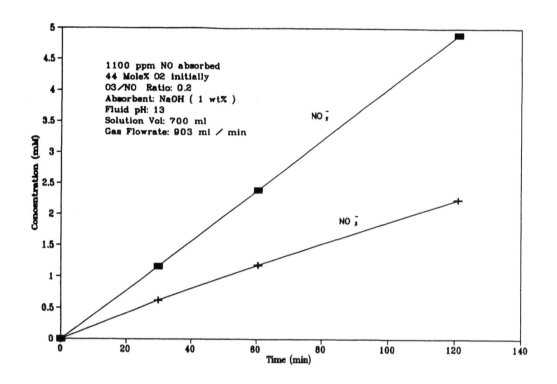

Figure 7 Concentration of NO_2^- as a Function of Time

ANALYSIS OF THE SCRUBBING SOLUTION

The contents of the spent NaOH solution in the bubbling reactor after NO ozonation experiments were analyzed by a spectrophotometric method (Hach DR/3000 Procedure). The ferrous sulfate method has been used as a measure of nitrite. By controlling the sample pH value, the nitrite present is deduced to nitrous oxide which reacts with the indicator to form a greenish brown color. Nitrates are not registered in the test. It was assumed that all the NO removed from the simulated flue gas can be recovered as a mixture of nitrite (NO_2^-) and nitrate (NO_3^-) [4,24]. Therefore, the mass balance of nitrogen (N) was studied in a closed reaction system and described as a mole concentration such as [NO_2^-], [NO_3^-] (mM). Results of this measurement for nitrite, involving 1105 ppm NO and gas flow rate of 903 ml/min at 22° C, are presented in Figure 14. Mole concentrations of NO_2^- and NO_3^- increase linearly with time. About two thirds of the removed NO were converted to [NO_2^-] and one third converted to [NO_3^-].

NO OXIDATION BY O$_2$ IN GAS AND LIQUID PHASES

The gas phase oxidation of NO can be expressed by the following stoichiometric equation:

$$2NO + O_2 \rightarrow 2NO_2 \tag{1}$$

TABLE I EXPERIMENTAL DATA FOR NO OXIDATION IN THE GAS PHASE

O_3/NO Ratio	NO Concentration in Outlet (ppm)	Percent NO Removal (%)
0.0	900	18.6
0.022	800	27.6
0.039	760	31.2
0.056	730	33.9
0.136	570	48.4
0.19	470	57.5
0.425	155	86.0

Influent concentration of NO = 1105 ppm; Initial O_2 concentration = 44.5 %; Gas flow rate = 903 ml/min; Residence time in experimental system (t_w) = 20 sec; Temperature = 25 °C; Pressure = 1 atm

The rate law is given by:

$$-\frac{d[NO]}{dt} = k_{GO2}[NO]^2[O_2] \tag{2}$$

where [NO] = concentration of NO, ppm;

$\quad$ [O_2] = concentration of O_2 , ppm;

$\quad$ k_{GO2} = rate constant for gas phase oxidation, (ppm^{-2} sec^{-1});

The experimental data are listed in TABLE I. The value of k_{GO2} was found to be 2.34 x 10^{-11} ppm^{-2} sec^{-1} (or 23.4 atm^{-2} sec^{-1}) at room temperature which is very close to the rate constant of 23.2 atm^{-2} sec^{-1} in literature [32].

$\quad$ The rate law NO oxidation in the liquid phase can be written as:

$$-\frac{d[NO]}{dt} = k_{LO2}[NO]^2[O_2] \tag{3}$$

The bubbling reactor in the arrangement of the experimental apparatus is set up very close to gas distributor. Before the reaction gas leaves the bubbling reactor, the total residence time was designed to be within two seconds; and NO oxidation continues in the gas phase until the gas mixture reaches the NO$_x$ analyzer. However, the residence time in this whole experimental system (t_w) can be estimated from the volume of tubing and bottles, gas flow rate, and rise velocity of a bubble. NO effluent concentration in the outlet of the bubbling

TABLE II EXPERIMENTAL DATA FOR NO OXIDATION IN THE LIQUID PHASE

O_3/NO Ratio	NO Concentration in Outlet (ppm)	Percent NO Removal (%)
0.0	670	39.4
0.022	500	54.8
0.039	415	62.4
0.056	380	65.6
0.136	100	91.0
0.19	< 7	99.4

Influent concentration of NO = 1105 ppm; Initial O_2 concentration = 44.5 % ; Gas flow rate = 903 ml/min; Residence time in experimental system (t_w) = 20 sec; Contact time in liquid phase (t_L) = 0.57 sec ; Absorbent = 10 wt% NaOH solution; pH value = 13; Temperature = 25 °C; Pressure = 1 atm

reactor can be easily calculated with k_{GO2}=2.34 x 10^{-11} ppm^{-2} sec^{-1}. Therefore, NO oxidation rate constant k_{LO2} can be estimated with different NO and O_2 influent concentration, contact time ,and NO effluent concentration in the liquid phase.

The experimental data are presented in TABLE II. The estimated k_{LO2} , 1.40 x 10^{-9} ppm^{-2} sec^{-1} , is approximately 60 times higher than that of gas phase oxidation when other experimental conditions (inlet gas concentration, temperature, and residence time) are kept the same.

NO OXIDATION BY O_3 IN GAS AND LIQUID PHASES

Experiments were conducted to compare NO ozonation rates with and without a bubbling reactor under otherwise identical concentration and operating conditions. These typical experimental results are reported in TABLES I and II, respectively. The NO oxidation rate by O_2/O_3 in both phases is presented in Figure 8 at different O_3/NO ratios. As shown in Figure 8, the ozonation rate of NO increases with the amount of ozone in both gas and liquid phases. In addition, the O_3/NO ratio in the liquid phase reaction was found to be half of the stoichiometric ratio required in the gas phase reaction. Figure 8 also shows that both oxidation (by O_2) and ozonation (by O_3) take place in our experiments and need to be accounted for in the following analysis.

The stoichiometric equation of gas phase NO ozonation can be expressed as:

$$NO + \alpha O_3 \xrightarrow{x_1} \text{Products such as } NO_2, O_2 \ldots \tag{4}$$

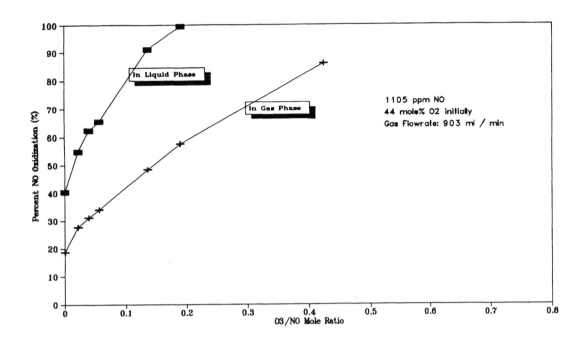

Figure 8 Comparison of NO Ozonation in gas and liquid phases (experimental values)

NO ozonation in the gas phase is assumed to follow a second order reaction [19].

The rate of NO oxidation, as indicated in the previous section, is assumed to be third-order. The stoichiometric equation is given by:

$$NO + 1/2O_2 \xrightarrow{\quad x_2 \quad} NO_2 \qquad (5)$$

Therefore, the combined rate of oxidation by O_2 and ozonation by O_3 in the gas phase is given by:

$$-\frac{d[NO]}{dt} = k_{GO3}[NO][O_3] + k_G[NO]^2[O_2] \qquad (6)$$

$$-\frac{d[O_3]}{dt} = k_{GO3}[NO][O_3] \qquad (7)$$

Where k_{GO3} = rate constant for ozonation in the gas phase, $ppm^{-1}\ sec^{-1}$;
$\qquad k_{GO2}$ = rate constant for oxidation in the gas phase, $ppm^{-2}\ sec^{-1}$;
$\qquad [NO]$ = concentration of NO, ppm
$\qquad [O_2]$ = concentration of O_2, ppm
$\qquad [O_3]$ = concentration of O_3, ppm
$\qquad x_1$ = conversion of NO reacting with O_3 in the gas phase;
$\qquad x_2$ = conversion of NO reacting with O_2 in the gas phase.

k_{GO3} and α values were obtained by fitting the model predictions to the experimental data. The best value of k_{GO3} was found to be 5.0×10^{-3} ppm^{-1} sec^{-1} with the stoichiometric NO to O$_3$ ratio being 1:3 ($\alpha = 1/3$). The reaction sequence is thus thought to be:

$$NO + O_3 \longrightarrow NO_2 + O_2 \tag{8}$$
$$2 NO + O_2 \longrightarrow 2 NO_2 \tag{9}$$

The calculated values from the model along with the experimental data at different O$_3$/NO ratios are presented in Figure 9.

The NO ozonation in the liquid phase can be expressed as follows:

$$NO + \alpha O_3 \xrightarrow{\ x_1'\ } \text{Products such as NO}_2, O_2, ... \tag{10}$$

$$NO + 1/2 O_2 \xrightarrow{\ x_2'\ } NO_2 \tag{11}$$

$$-\frac{d[NO]}{dt} = k_{LO3}[NO][O_3] + k_L[NO]^2[O_2] \tag{12}$$

$$-\frac{d[O_3]}{dt} = k_{LO3}[NO][O_3] \tag{13}$$

Where k_{LO3} = apparent rate constant for ozonation in the liquid phase, ppm^{-1} sec^{-1};

k_{LO2} = apparent rate constant for oxidation in the liquid phase, ppm^{-2} sec^{-1};

x_1' = conversion of NO reacting with O$_3$ in the liquid phase;

x_2' = conversion of NO reacting with O$_2$ in the liquid phase.

The distance between the mixing point of gas streams and the entrance into the liquid phase is very short in our experiments (about 0.43 second residence time). Besides, the gas flow is laminar before entering the bubbling reactor (Re=0.00004). Assuming that gas phase reaction can be neglected due to poor mixing initial NO concentrations and O$_3$/O$_2$ to NO ratios were used in this calculation. In order to compare with measured values, NO concentrations obtained from the NO$_x$ analyzer have been corrected to the NO concentrations at the outlet of the bubbling reactor.

The calculated values are presented along with the experimental data in Figure 10. The best value of k_{LO3} was found to be 0.6 ppm^{-1} sec^{-1} with the stoichiometric NO to O$_3$ ratio being 1:6 ($\alpha = 1/6$). These apparent rate constants indicate that the rate of ozonation in the liquid phase is more than 100 times faster than that in the gas phase under otherwise identical experimental conditions.

DISCUSSION

The significant removal rate of NO at a very low O$_3$/NO ratio in high alkaline solution may be explained as follows. The oxidation of NO to NO$_2$ (by O$_2$) may be effected by ozone in the gas phase. The NO_2 thus produced can either react with another molecule of NO to form N_2O_3 or dimerizes to form N_2O_4 [24]. Both N_2O_3 and N_2O_4 are much more soluble and more reactive in aqueous solution compared to NO, and their dissolution in aqueous solution leads to the formation of nitrite and nitrate:

$$3 \, NO + O_3 \rightarrow 3 \, NO_2 \tag{14}$$
$$3 \, NO + 3 \, NO_2 + 6 \, OH^- \rightarrow 6 \, NO_2^- + 3H_2O \tag{15}$$
$$2 \, NO_2^- + O_2 \rightarrow 2 \, NO_3^- \tag{16}$$

or

$$O_3 + 6NO + O_2 + 6 \, OH^- \rightarrow 4 \, NO_2^- + 2 \, NO_3^- + 3H_2O \tag{17}$$

Our studies demonstrate that liquid phase ozonation is faster in kinetics and consumes less ozone per mole of nitric oxide than gas phase ozonation. The implication is that ozone can be bubbled through a scrubbing liquor along with the flue gas to achieve NO_x removal more economically than gas phase ozonation. The larger apparent rate constants K_{LO3} and K_{LO2} indicate that NO, the main insoluble part of NO_x, can be removed with a faster kinetics or smaller reactor volume. The high stoichiometric O_3 to NO ratio means less O_3 is required for NO_x removal in liquid phase operations. The faster kinetics can be attributed to the enhancement resulted from the chemisorption of NO. Since O_2 is also a good oxidizing agent, it may be beneficial to bubble air through the reactor to achieve simultaneous oxidation and ozonation of NO.

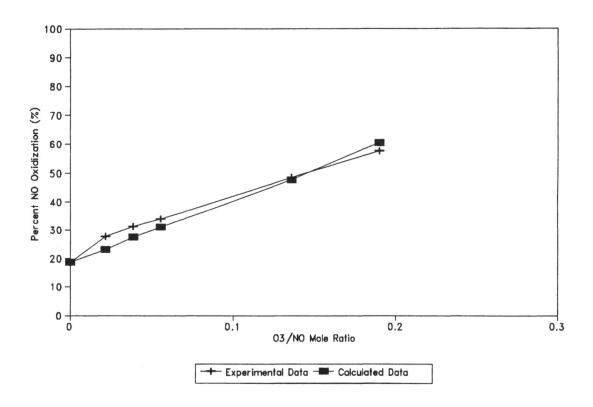

Figure 9 Experimental and Estimated Results for NO Oxidation in the Gas Phase

CONCLUSIONS

1. More than 99 % removal can be achieved at pH=13 with O_3/NO ratio as low as 0.2 in a bubbling reactor. NO ozonation in both gas and liquid phases can be adequately described by second-order rate laws. The apparent rate constant k_{LO3} was found to be 0.6 ppm^{-1} sec^{-1} with the stoichiometric NO to O_3 ratio in the liquid phase being 1:6. The value of k_{GO3} was found to be 0.005 ppm^{-1} sec^{-1} with the stoichiometric NO to O_3 ratio in the gas phase being 1:3.

2. Oxygen is also a good oxidizing agent under bubbling reactor conditions. NO oxidation rate laws in both gas and liquid phases are second order in NO and first order in O_2. The estimated k_{LO2} value for liquid phase is 1.4×10^{-9} ppm^{-2} sec^{-1}, much higher than the gas phase k_{GO2} value of 2.34×10^{-11} ppm^{-2} sec^{-1}.

3. The effect of pH on NO removal by O_3 is very pronounced as the pH values vary from 1 to 13 using various reagents (HCl, HNO_3, buffer solution, $NaHCO_3$, Na_2CO_3, and NaOH). A higher pH favors the NO removal.

4. The effect of scrubbing liquid temperature on NO removal from 22 °C to 70 °C is significant in ozonation. A lower operating temperature favors the removal of NO.

5. Ozone is found to be more effective at lower NO concentrations, where a better utilization of O_3 was observed.

ACKNOWLEDGMENT

Financial support from U.S. EPA/ Gulf Coast Hazardous Substance Research Center under Grant No. 111LUB0256 is gratefully acknowledged.

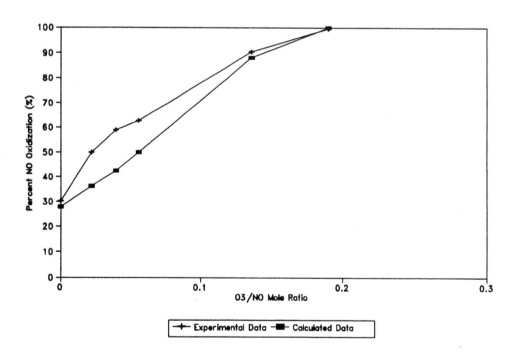

Figure 10 Experimental and Estimated Results for NO Oxidation in the liquid Phase

REFERENCES

1. Amano, M. and H. Tsugawa, 1977. "Pilot plant for a wet simultaneous SOx and NOx removal process using lime and limestone scrubbing," IHI Eng. Rev., **10(4)**, 72-77.

2. Buck, M., Clucas, J., McDonogh C., and Woods, S., "NO_x Removal in the Stainless Steel Pickling Industry with Hydrogen Peroxide." Proceedings of the first International Symposium on Chemical Oxidation: Technologies for the Nineties, Ed. Eckenfelder, W. W., A. R. Bowers and J. A. Roth, Feb. 20-22, 1991, Nashville, Tennessee.

3. Chang, S. G. and G. C. Lee, "LBL PhoNOX Process for combined Removal of SO2 and NOx from Flue Gas," 1991 AIChE Summer National Meeting, Pittsburgh, PA August 18-21, 1991.

4. Carta, G., 1984. "Role of HNO_2 in the absorption of Nitrogen Oxides in Alkaline Solutions." Ind. Eng. Chem. Fundam. 23, 260-264.

5. Chang, S. G. and D. K. Liu, 1990. "Removal of Nitrogen and Sulphur Oxides from Waste Gas Using A Phosphorus/Alkali Emulsion." Nature, 343(11), 151.

6. Clearwater, S. W. and J. M. Scanlon, 1991. "Legal Incentives for Minimizing Waste," *Environmental Progress*, **10(3)**, 169.

7. Cobb, D., L. Glatch, J. Ruud, and S. Snyder, 1991. "Application of Selective Catalytic Reduction (SCR) Technology for NOx Reduction from Refinery Combustion Sources," *Environmental Progress*, **10(1)**, 49.

8. Colannino, J, "Low-Cost Techniques Reduce Boiler NO_x," *Chem. Eng.*, Feb. 1993, 100-106.

9. Cooper, C.D. and F.C. Alley, 1986. "Air pollution control: a design approach", (PWS Engineering, Boston, MA.

10. Elsom, D., 1987. "Atmospheric pollution", Basil Blackwell, New York, New York.

11. Garg, A,,"Trimming NO_x from Furnaces," *Chem. Eng.*, Nov. 1992, 122-129.

12. Gilbert, B. R., Y. Kogawa, M. Sawahata, and A. Kumagai, 1987. "Commercial Status of the Chiyoda Thoroughbred 121 (CT-121) Flue Gas Desulfurization Process," *Proceedings Am. Power Conf.*, San Francisco, CA., **49**, 51-7.

13. Glaze, W.H. and J.W. Kang, J. AWWA, May (1987), 57-63.

14. Haggin, J., "Energy Policy: Coal." Chem. & Eng. News, June 17, 32 (1991).

15. Hallstrom, R. A. U. and D. W. Johnson, "From Acid Rain to Wallboard: Reducing Secondary Pollution in Flue Gas Desulfurization," 1991 AIChE Summer National Meeting, Pittsburgh, PA August 18-21, 1991.

16. Hoigne, J. and H. Bader, 1976. "The Role of Hydroxyl Radical Reactions in Ozonation Processes in Aqueous Solution." Water Research, 10, 377-386.

17. Idemura, H., "Simultaneous SO2 and NOx removal process for flue gas," Chemical Economy and Engineering Review, August 1974, **6(8)**, 22-26.

18. Jethani, K.R.; N.J. Suchak and J.B. Joshi, 1990. " Selection of reactive solvent for pollution abatement of NOx," Gas Separation & Purification, 4, 8-28.

19. Johnston, H.S., and J.H. Vrosby, 1954. "Kinetics of the Fast Gas Phase Reaction Between O3 and NO." J. Chem. Phys., 22, 689.

20. Kim, S.S. and C.J. Drummond, "Advanced flue gas cleanup technology", Paper presented at the Fifth U.S./Korean Joint Work- shop on Coal Utilization

Technology, New Orleans, Louisiana, October 24-26, 1988.

21. Kuo, C. H. 1982. "Mass Transfer in Ozone Absorption." Envir. Prog., 1(3), 189-195.

22. Li, K.Y., Thesis, Mississippi State Univ., Dec., 1977.

23. Littlejohn, D. and S. G. Chang, 1991. *Energy & Fuels*, 5, 249.

24. Liu, D. K.; D. X. Shen; and S. G. Chang, 1991. "Removal of NO_x/SO_2 from Flue Gas Using Aqueous Emulsions of Yellow Phosphorus and Alkali," *Environ. Sci. Technol.*, 25(1), 55.

25. Lowe, P.A., W. Ellison, and L. Radak, "Assessment of Japanese SCR Technology for Oil-fired Boilers and Its Applicability in the USA," *Proceedings of the EPRI/EPA 1989 Joint Symposium on Stationary Combustion NOx Control*, March 6-9, 1989, San Francisco, CA.

26. Moyeda, D. K., B. A. Folsom, T. M. Sommer, Q. H. Nguyen, D. K. Hartsock, and J. C. Opatrny, "Demonstration of Combined NOx and SO2 Emission Control Technologies involving Gas Reburning," 1991 AIChE Summer National Meeting, Pittsburgh, PA August 18-21, 1991.

27. Noack, M. G. and S. A. Iacoviello, "The Chemistry of Chlorine Dioxide in Industrial and Wastewater Treatment Applications." presented at Chemical Oxidation Technologies for the Nineties, Feb. 20-22, 1991, Nashville, Tennessee.

28. Overath, H., 1978. "Use of Hydrogen Peroxide in Water Treatment," *Proc. Oxidation Techniques in Drinking Water*, Karlsruhe, FRG, CCMS 111, USEPA, 544-555.

29. Precision Scientific Inc., Chicago, IL, Instruction Manual. TS-63111 AT-8.

30. Rice, R. G. and Wilkes, J. F., "Ozone Chemistry Applied to Cooling Tower Water Treatment," *Proc. 2nd Int. Sym. on Chemical Oxidation Technology for The Nineties*, Vanderbilt University, Nashville, Tennessee, Feb. 19-21, 1992.

31. Rodier, J, 1975. Analysis of Water. John Wiley & Sons, Inc., New York.

32. Sherwood, T.K. and R.L. Pigford, 1952. Absorption and Extraction. McGraw-Hill, New York, 1952.

33. Theodore, L. and A.J. Buonicore, 1988. "Air pollution control equipment", 1, CRC Press. Boca Raton, Florida, 1988.

34. Wallach, D., Beaumont Enterprise, June 27, 1991.

35. Yeh, J. T., C. J. Drummond and J. I. Joubert, 1987. "Process Simulation of the Fluidized-Bed Copper Oxide Process Sulfation Reaction." Envir. Prog., 6(2), 44.

36. Yeh, J. J., W. T. Ma, H. W. Pennline, J. L. Haslbeck and F. N. Gromicko, "Integrated testing of the NOXSO process: simultaneous removal of SO2 and NOx from flue gas", paper presented at AIChE Spring National Meeting, March 18-22, 1990, Orlando, Florida.

JAMES R. BOLTON
STEPHEN R. CATER
ALI SAFARZEDEH-AMIRI

Advanced Oxidation Technologies for the Photodegradation of Organic Pollutants in Water

ABSTRACT

Advanced oxidation processes (AOP) can be used for the treatment of a variety of pollutants in water. There are several type of processes available and the choice of the process depends on the nature of the water and pollutants to be treated. For easy to treat waters, direct photolysis and UV/peroxide offer cost effective treatment solutions. For difficult to treat waters, several patented or proprietary options are available. These options can dramatically reduce the capital and operating costs of treatment systems and make AOP the most cost effective treatment.

INTRODUCTION

The use of advanced oxidation processes (AOP) to remediate contaminated water usually relies on the generation of hydroxyl radicals to oxidize the pollutants. A common method for production of hydroxyl radicals involves photolysis of hydrogen peroxide with ultraviolet light. This technique is generally effective in the treatment of ground waters contaminated with ppb or ppm levels of pollutants. There are cases however when photolysis of hydrogen peroxide does not offer an economical solution. These include the treatment of refractory pollutants or the treatment of heavily contaminated and/or highly UV absorbing waters. Refractory pollutants, such as haloalkanes, react more slowly with hydroxyl radicals and hence their treatment becomes more costly when compared to treatment of non-refractory pollutants such as trichloroethylene or benzene. Heavily contaminated waters treated with UV/peroxide have higher treatment costs due to the high background UV absorbance, which blocks the UV light from photolyzing the peroxide. Also the high loadings of contaminants typically result in large amounts of hydrogen peroxide being required for satisfactory treatment.

Over the past several years, we have developed photocatalysts and process improvements to increase the size of the market where AOP is cost effective. This paper will outline some of these developments and our approach to the treatment of contaminated water.

James R. Bolton, Stephen R. Cater and Ali Safarzedeh-Amiri, Solarchem Environmental Systems, 130 Royal Crest Court, Markham, Ontario, Canada L3R 0A1

RESULTS

Each type of water requires a determination of its characteristics to find the most cost effective AOP treatment option. Each water is analyzed for parameters such as UV absorbance, chloride, alkalinity and chemical oxygen demand (COD) to determine the level of background contamination. These influence, along with the nature and concentration of the particular pollutants, the preferred treatment option. These options are discussed below.

Rayox® Direct Photolysis:

In some cases it is possible to treat contaminated water by direct UV photolysis of the contaminant. A good example is the treatment of water contaminated with N-nitrosodimethylamine (NDMA). NDMA has a strong absorption band in the UV region with the λ_{max} at 228 nm and a quantum yield of dissociation of approximately unity. Photolysis, therefore, results in the degradation of the molecule. This treatment option requires a UV lamp with high output in the 200 nm - 240 nm range. **Rayox®** treatment systems for this application are very simple with very low operating costs. This application is only applicable in waters of low UV absorbance so as to allow for absorption of the UV light by the target pollutant.

Rayox® UV/Peroxide:

In North America, the combination of UV/peroxide represents the most commonly utilized AOP technology. In this case, hydrogen peroxide absorbs UV light and photodissociates into hydroxyl radicals. Since peroxide is a weak absorber, the process is best suited to the treatment of groundwaters contaminated with pollutants in the ppb to <100 ppm range.

Rayox®-F:

This is a patented process (U.S. Patent #5,043,080) involving the use of transition metal ions to increase the rates of destruction of pollutants. Iron salts are usually added in the low ppm range along with hydrogen peroxide at an acidic pH. The process (commonly referred to as the photo-Fenton's reaction) is best suited to treatment of aromatic and olefinic pollutants. Since the iron (III) absorbs light and generates iron (II) plus hydroxyl radical and Fenton's reaction also generates hydroxyl radicals, the **Rayox®-F** process can be more cost effective in treating waters of higher background absorbance than is simple UV/peroxide. Treatment rates can be increased up to 2 to 5 fold, when compared to simple UV/peroxide.

Rayox®-O UV/Ozone:

Photolysis of ozone in water generates hydrogen peroxide, which in turn photolyses to hydroxyl radicals or reacts with ozone to give hydroxyl radicals, and thus is a rather inefficient way to produce these radicals. However, there are cases where the use of ozone offers a much more cost effective approach than UV/peroxide. The treatment of trinitrotoluene (TNT) in water is one such case. Hydroxyl radical attack on TNT generates trinitrobenzene (TNB) as a by-product which is quite refractory to further degradation by hydroxyl radicals. This means that the overall cost of treatment of TNT in water is largely determined by the discharge level of TNB.

The use of UV/ozone is a more effective means of treating TNT contaminated water. In this case TNB is not created and the treatment of TNT with UV/ozone is consequently very cost effective. Figure 1 compares the treatment rates (as a function of total applied energy dose) of TNB by Uv/peroxide and by UV/ozone and shows the benefit of UV/ozone treatment.

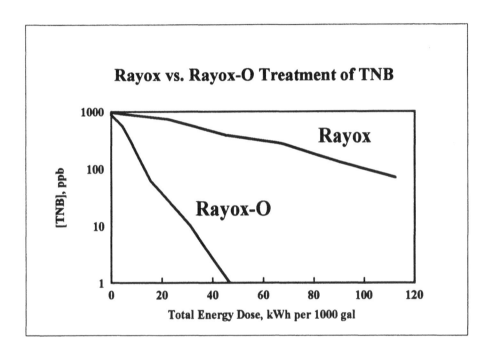

Figure 1

Rayox®-R:

This is a patented photo-reduction process (U.S. Patent #5,258,124) for the treatment of haloalkanes. These compounds, which include such pollutants as trichloroethane, chloroform, carbon tetrachloride and freons, are refractory to oxidation by hydroxyl radicals. The process involves the photochemical generation of hydrated electrons (e_{aq}^-), which are powerful reducing agents and are highly reactive with halogenated compounds. The hydrated electrons in **Rayox®-R** are produced by the photolysis of iodide ion:

$$I^- + h\nu \rightarrow e_{aq}^- + I\bullet$$

The hydrated electrons attack the halogenated compound and progressively eliminate halide ion from the contaminant.

Iodide has a strong absorption band centered at 222 nm, so the strong emission below 240 nm of Solarchem lamps is utilized to effect the photolysis. A reducing agent is added to recycle the I• back to iodide. Thus, only catalytic amounts of iodide are required.

The treatment rates for the haloalkanes are increased from 2 to 10 fold over those using hydroxyl radical based processes, thus offering a cost effective solution for these pollutants. Figure 2 compares the treatment of chloroform by UV/peroxide and by the **Rayox®-R** process and demonstrates the improvement possible with the **Rayox®-R** process.

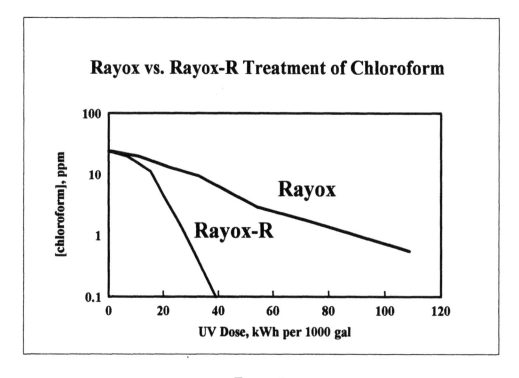

Figure 2

Rayox®-A:

This is a patented process (U.S. Patent #5,266,214) involving the photolysis of iron (III) oxalate photocatalyst. This compound absorbs light strongly between 200 nm and 500 nm. Photolysis of the iron oxalate generates iron (II) which then undergoes a Fenton's reaction with hydrogen peroxide to yield hydroxyl radicals. Figure 3 compares the absorbance spectrum of iron (III) oxalate with that of hydrogen peroxide. Since the photocatalyst absorbs light much more strongly and over a much broader wavelength range than peroxide, the process can be applied to waters that are much more heavily contaminated. Treatment efficiencies are improved by up to 10 fold or more as compared to UV/peroxide. Figure 4 compares the treatment of a process water contaminated with benzene, toluene, ethylbenzene and xylene (BTEX) by UV/peroxide and by **Rayox®-A**, demonstrating the enhancement possibility. The process can be used for the treatment of waters with hundreds of ppm of contaminants and up to a few thousand ppm of COD. These waters cannot be economically treated by other UV based AOP processes. Thus, the range of applicability of AOP can be greatly increased by the use of the **Rayox®-A** process.

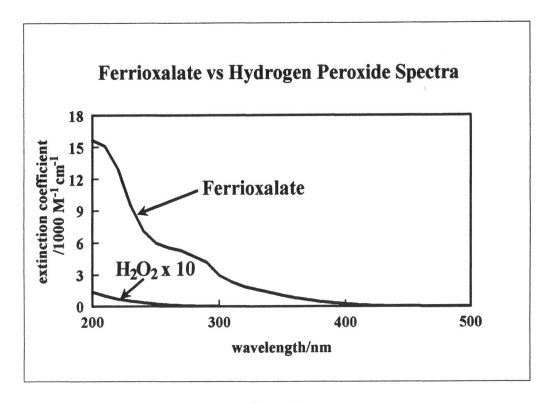

Figure 3

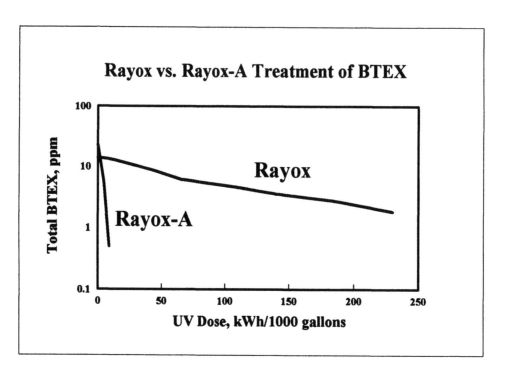

Figure 4

Solaqua®:

This process uses the same chemistry as **Rayox®-A** but employs sunlight in place of discharge lamps. It is thus a homogeneous, sunlight driven decontamination process for water. The photocatalyst absorbs about 18% of the solar spectrum as compared to 3% absorbed by titanium dioxide, as demonstrated in Figure 5. In addition, the quantum yield of hydroxyl radical production is approximately unity, compared to about 0.05 for titanium dioxide. The **Solaqua®** process offers treatment efficiencies of 50 to 100 times that of titanium dioxide. Figure 6 compares the treatment of toluene with **Solaqua®** and titanium dioxide and shows the improved treatment possible with the **Solaqua®** process.

CONCLUSIONS:

The type of AOP to use for treatment of contaminated water depends on the nature and concentration of the contaminant and on the nature of the water. We have developed new photocatalysts and new processes to deal with the treatment of difficult to treat waters. These processes include a new photoreduction process, **Rayox®-R**, for the treatment of halogenated alkanes, **Rayox®-A** for the treatment of highly contaminated streams and **Solaqua®**, which is a homogeneous sunlight based treatment process.

114

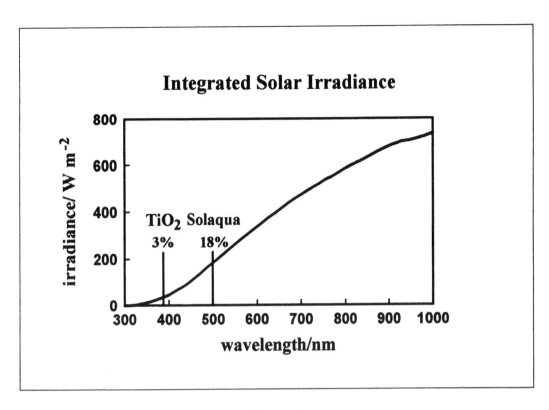

Figure 5

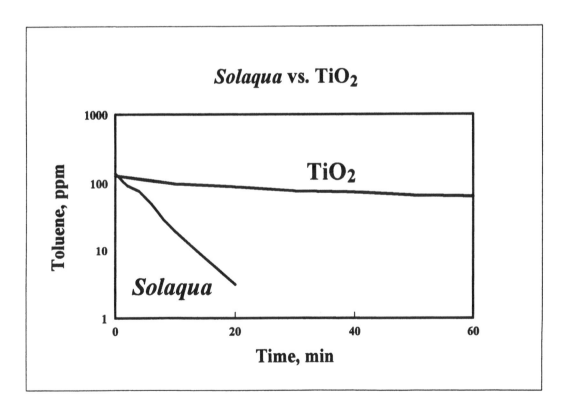

Figure 6

115

FRANK Y. C. HUANG
K. Y. LI
C. C. LIU
Z. F. LIU

Treatment of Groundwater Contaminated with Chlorinated Hydrocarbons

ABSTRACT

The groundwater aquifer underneath a chemical manufacturing plant in Southeast Texas has been contaminated with the leachate from its landfill. There are 17 major chlorinated hydrocarbon contaminants found in the groundwater with concentrations ranging from 1 ppm to 1200 ppm. An air-stripping unit followed by a thermal catalytic oxidation unit is currently operating on-site to remove all of the chlorinated compounds from the contaminated groundwater. Because one of the contaminants, Bis(2-chloroethyl)ether (DCEE), has a fairly low Henry's law constant, a high air flow rate is employed in the stripping unit in order to improve the overall stripping efficiency. This high air flow rate has caused maintenance problems in the air-stripping unit and a higher energy consumption in the thermal catalytic oxidation unit. Also, the effluent from the air stripping unit still contains a fair amount of DCEE which may require further treatment before being discharged into the receiving stream in the event of a tighter effluent permit.

A photo-chemical oxidation unit was set up to study the feasibility of using UV/H_2O_2 to destroy DCEE. Kinetics and factors which may affect the effectiveness of DCEE oxidation were investigated. The oxidation of 1,2-Dichloroethane (DCA), 1,1,2-Trichloroethane (TCA), and 1,1,2,2-Tetrachloroethane (TetCA) were also studied because they have the most significant concentrations in the contaminated groundwater. The information gathered from the oxidation of these three compounds will be used to determine whether the UV/peroxidation process can be used as an alternative treatment process for the contaminated groundwater.

The stoichiometric ratio of DCEE to H_2O_2 is 1:10 for the complete oxidation of DCEE. Under this condition, over 99% of the DCEE at an initial concentration of 220 ppm was oxidized after 10 minutes of irradiation. Several volatile oxidation intermediates were observed at the early stage of the oxidation process; however, nearly all of the by-products were not detected after 30 minutes of irradiation. No by-products of high molecular weights were ever observed even at very low H_2O_2 doses

The results from the oxidation of 1,2-DCA, 1,1,2-TCA, and 1,1,2,2-TetCA indicated that these compounds are much more recalcitrant to the UV/peroxidation process than DCEE. Stoichiometic amounts of H_2O_2 for the complete oxidation of chloroethanes were always required in order to prevent the formation of high molecular weight compounds.

Frank Huang, Gulf Coast Hazardous Substance Research Center, P.O. Box 10613, Beaumont, TX 77710
K.Y. Li, C.C. Liu, and Z. F. Liu, Dept. of Chemical Engineering, Lamar University, Beaumont, TX 77710

INTRODUCTION

The groundwater aquifer underneath a chemical manufacturing plant in Southeast Texas has been contaminated with the leachate from its landfill. There are 17 major chlorinated hydrocarbon contaminants found in the groundwater. Their concentrations and Henry's law constants are shown in Table 1. High levels of salinity and hardness were also observed in the contaminated groundwater as shown in Table 2. An air-stripping unit followed by a thermal catalytic oxidation unit is currently being operated on-site to remove all of the chlorinated compounds from the contaminated groundwater. Figure 1 gives a general layout of the existing process. The current operating air flow rate of the air-stripping unit was designed to provide adequate stripping for DCEE which has a Henry's law constant that is much lower than that of any other chlorinated contaminants in the groundwater. The off-gas from the air stripping unit is channeled directly into the thermal catalytic oxidation unit for further treatment and the treated groundwater is discharged directly into a bayou nearby [1].

TABLE I - MAJOR CONTAMINANTS FOUND IN GROUNDWATER

Component	Concentration (ppm)	Henry's law constant (atm)
1,2-Dichloroethane	1200	61.16
1,1,2-Trichloroethane	400	47.04
1,1,2,2-Tetrachloroethane	200	24.02
Bis(2-chloroethyl)ether	200	1.2
Methylene Chloride	100	177.37
Perchloroethylene	100	1595.8
1,1-Dichloroethane	50	303.03
Chloroform	40	188.49
Trichloroethylene	24	650.54
1,2-Dichloroethene	22	295.8
1,1-Dichloroethene	5	834.03
1,1,1-Trichloroethane	5	273.56
Carbon Tetrachloride	7	1679.17
Propylene Dichloride	10	156.80
Hexachloroethane	5	547.68
Hexachlorobutadiene	15	572.7
Hexachlorobenzene	1	94.52

TABLE II - GENERAL CHARACTERISTICS OF GROUNDWATER

Major Components	
Total Alkalinity as $CaCO_3$	405 ppm
pH	6.5
Chloride	48,000 ppm
Sulfate	2,330 ppm
Calcium	970 ppm
Magnesium	960 ppm
Sodium	18,000 ppm
Iron	3 ppm

Based on computer simulations, the current air flow rate operated in the air stripping unit is about eight times higher than the calculated flow rate if DCEE is excluded. This excessive air flow rate has caused maintenance problems in the air-stripping unit and a higher energy consumption in the thermal catalytic oxidation unit. Also, the effluent from the air stripping unit still contains a fair amount of DCEE which may require further treatment before being discharged into the receiving stream in the event of a tighter effluent permit [2]. A photo-chemical oxidation unit was set up to evaluate its effectiveness on the oxidation of DCEE.

With the Clean Air Act Amendment which took effect in 1992, more stringent regulations are expected to be promulgated on the emissions of volatile organic compounds [3]. Therefore, the oxidation of the three chlorinated contaminants with the most significant concentrations in the groundwater was also studied in order to gather preliminary information on whether the photo-chemical oxidation unit can be used as an alternative treatment process for the contaminated groundwater. The three chlorinated contaminants studied were 1,2-DCA, 1,1,2-TCA, and 1,1,2,2-TetCA.

Figure 1. A general layout of the existing process

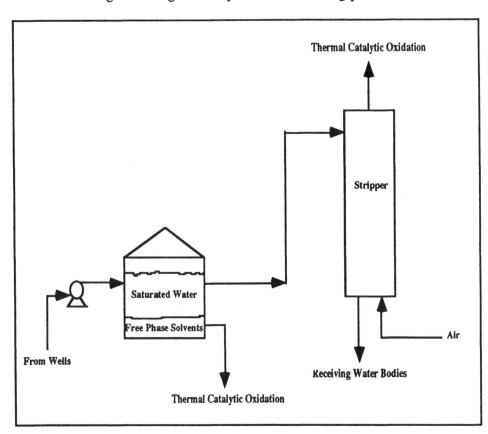

Although a lot of studies have been done on the photo-chemical oxidation of organics [4-7], very little data is available about the oxidation of DCEE in a UV/peroxidation process. However, Kim and others have demonstrated successfully the cost-effectiveness of UV-peroxide degradation on methyl *tert*-butyl ether, a compound with a similar chemical structure to DCEE [8-9]. Chlorinated ethanes, on the other hand, have been studied more thoroughly and shown to be amenable to UV/peroxidation treatment [10-11]. Besides the type of contaminants to be treated, several other factors will also affect the effectiveness of a UV/peroxidation process. Anions such as bicarbonate and carbonate, can serve as scavengers for the oxidant. Metals present in their reduced states, such as ferrous iron, are likely to be oxidized to less soluble forms, which can cause the scaling problems on UV lamps. Operational and environmental parameters such as hydraulic retention time in the photo-chemical reactor, hydrogen peroxide dose, UV lamp intensity, pH, and temperature will also influence the performance of a UV/peroxidation system [12].

EXPERIMENTAL METHODS

A medium-pressure, mercury, immersion lamp with a 1-liter borosilicate glass reaction vessel (ACE Glass Inc.) was used in the study. Figure 2 shows the schematic diagram of the photo-chemical oxidation unit. Of the total energy radiated from the lamp (175.8 Watts), approximately 40-48% is in the ultraviolet portion of the spectrum, 40-43% in the visible, and the balance in the infrared. All the experiments were conducted in a complete-mixed, batch mode. The water temperature and pH were maintained at about 23 °C (through the recirculation of cooling water) and 7 (through the addition of a phosphate buffer solution) respectively, unless otherwise specified. The concentration of DCEE studied was about 200 ppm which represented its concentration in the contaminated groundwater. Oxidation reactions of different DCEE to H_2O_2 ratios (molar ratio) were examined first and the concentrations of DCEE and H_2O_2 were monitored along the course of the reactions. The concentrations of DCEE were analyzed using a gas chromatograph (GC) (Varian model 4500) equipped with a purge-and-trap concentrator (Tekmar model LSC 2000), a flame-ionization detector, and a 30-meter DB™-1701 Microbore® column (J&W Scientific). The concentrations of H_2O_2 were measured using a modified titration method [2]. A gas chromatograph (HP model 5890 Series II) equipped with a mass spectroscopy detector (HP model 5971 Series MSD), a sample concentrator (OI model 4460), and a HP Ultra-2 capillary column was used to identify oxidation by-products in all of the reactions.

Oxidation kinetics of the UV/H_2O_2 process was studied using the initial rate method [13]. In this method, the initial concentration of DCEE was first kept constant at about 200 ppm and its initial oxidation rates (time≈ 0) were obtained for different H_2O_2 doses. Then, the initial concentration of H_2O_2 was kept constant and the initial oxidation rates of DCEE were obtained for different DCEE doses. Based on these two sets of data, empirical rate equations of the UV/peroxidation of DCEE and H_2O_2 can be estimated.

Based on the general characteristics of the contaminated groundwater, several other environmental and operational factors which may affect the efficiency of the UV/H_2O_2 process on the oxidation of DCEE were also investigated. The effectiveness of different portions of the spectrum emitted by the lamp was studied using filter sleeves which surrounded the light source to absorb the spectrum of certain wavelength ranges. The Corex sleeve was used to block about 58% of the spectrum in the far UV range (220 nm-280 nm) and the Pyrex sleeve was used to block 100% of the Far UV spectrum as well as about 65% of the middle UV spectrum (280 nm-320 nm). The impact of alkalinity (1000 mg/L as $CaCO_3$), high salt content (55g/L of NaCl), pH variations (no addition of the phosphate buffer solution), and lower temperature (at 11 °C) on the oxidation process were also investigated in order to further understand the possible inhibition factors of the oxidation process.

Only preliminary data was gathered on the UV peroxidation of 1,2-DCA, 1,1,2-TCA, and 1,1,2,2-TetCA. Their targeted concentrations were 1200 ppm, 400 ppm, and 200 ppm respectively. Different H_2O_2 to chlorinated compound ratios were explored and the concentrations of the chlorinated compound, H_2O_2, as well as oxidation by-products were monitored along the course of the reactions.

Figure 2. A schematic diagram of the photo-chemical oxidation unit

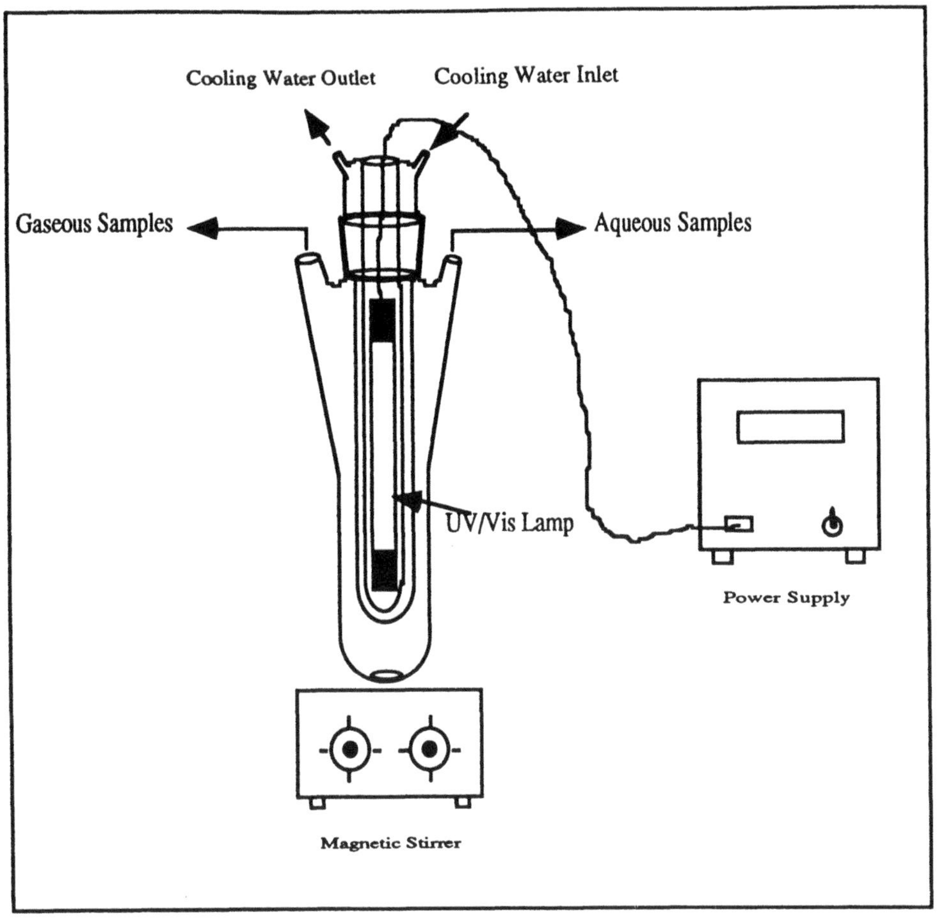

OXIDATION OF DCEE

H_2O_2 DOSES

The concentrations of DCEE with respect to the reaction time under different DCEE to H_2O_2 ratios are shown in Figures 3 and 4. The stoichiometric ratio of DCEE to H_2O_2 is 1:10 for the complete oxidation of DCEE. Under this condition, over 99% of the DCEE at an initial concentration of 220 ppm was oxidized after 10 minutes of irradiation. Chloromethane, vinyl chloride, chloroethane, and ethyl chloroacetate were observed concurrently as the oxidation intermediates and all appeared to be in the concentration range of several ppb or below. Two other unidentified compounds, both are more volatile than DCEE, were also observed at trace levels in the mass spectrum. Nearly all of the by-products were not detected after 30 minutes of irradiation, . At lower H_2O_2 doses, DCEE (at an initial concentration of 220 ppm) also was not detected after 30 minutes of irradiation (except at H_2O_2/DCEE=0.9); however, further oxidation of the by-products has been observed to be less.

Figure 3. DCEE vs Time

(initial DCEE conc. = constant)

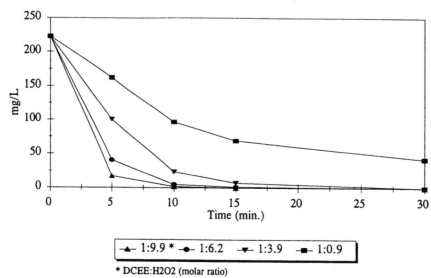

* DCEE:H2O2 (molar ratio)

Legend: 1:9.9 *   1:6.2   1:3.9   1:0.9

Figure 4. DCEE vs Time

(initial H2O2 conc. = 290 mg/L)

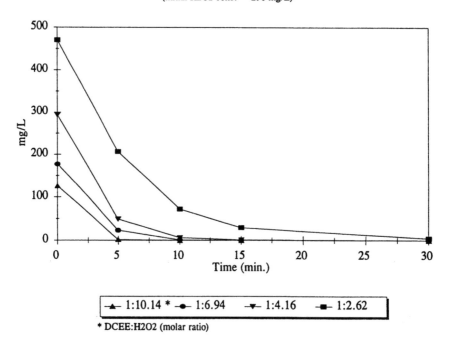

* DCEE:H2O2 (molar ratio)

Legend: 1:10.14 *   1:6.94   1:4.16   1:2.62

OXIDATION KINETICS

Based on the initial rate method, two empirical rate equations for the UV/H$_2$O$_2$ process on the oxidation of DCEE at 23 °C and pH=7 were obtained from the data shown in Figures 3-4 and are listed below.

$$-\frac{d(DCEE)}{dt} = 0.065(DCEE)^{0.675}(H_2O_2)^{0.716} \tag{1}$$

$$-\frac{d(H_2O_2)}{dt} = 0.0196(DCEE)^{-0.162}(H_2O_2)^{1.565} \tag{2}$$

$DCEE$ = the concentration of DCEE in the aqueous phase (mM)

H_2O_2 = the concentration of H$_2$O$_2$ in the aqueous phase (mM)

t = time (minute)

A computer model of the above empirical rate equations was built using a simulation software, TUTSIM™ (TUTSIM Products, CA). Figure 5 shows the comparison between experimental data and the model predictions at different DCEE to H$_2$O$_2$ ratios. The model matches well with the experimental data at higher H$_2$O$_2$ doses. However, over prediction of the DCEE oxidation rate may occur when the H$_2$O$_2$ doses are much lower than the stoichiometric amount required for the complete oxidation of DCEE.

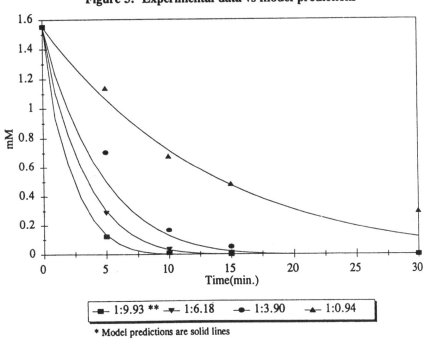

Figure 5. Experimental data vs model predictions *

- ▪ 1:9.93 **   ▼ 1:6.18   ● 1:3.90   ▲ 1:0.94

* Model predictions are solid lines
** DCEE:H2O2 (molar ratio)

The effectiveness of different portions of the spectrum on the oxidation of DCEE is shown in Figure 6. Since the DCEE oxidation rate with the Corex sleeve was comparable to that with Quartz only, it suggested that the spectrum very close to 200 nm may not be as critical as the rest of the far UV spectrum. From the experiment with the Pyrex sleeve, it also indicated that the spectrum at the near UV, visible, and infrared ranges was not significant in promoting the oxidation process. These information suggested that a low-pressure mercury lamp, which emits a spectrum primarily around 254 nm, may be sufficient to provide the similar DCEE removal efficiency of the medium-pressure mercury lamp. Also minimum cooling is required for the low-pressure mercury lamp which will simplify the operation and maintenance of the photo-chemical oxidation unit.

The effects of temperature, salinity, pH, and alkalinity on the oxidation process are shown in Figures 7-10. No significant differences were observed between the DCEE oxidation rate at 11 °C and that at 23 °C. This result indicated that the design of the oxidation process may not be constrained by the seasonal temperature variation of the air-stripped groundwater. The addition of buffer solutions to maintain the pH of the groundwater close to 7 may be required in the oxidation process because some reduction of the DCEE oxidation rate was observed in the case where pH was not controlled during the oxidation process. However, if sufficient contact time was provided (more than 15 minutes) in the oxidation reactor, the influence of pH variations on the oxidation process could be minimized. No hindrance has been observed on the oxidation process at the salinity concentration (55 g/L of NaCl) equivalent to that of the contaminated groundwater. Although bicarbonate ions have been shown to serve as scavengers for hydroxyl free radicals[10], no significant interference has been observed at the concentration of 1000 mg/L (as $CaCO_3$) in our studies.

Figure 6. Effect of absorption sleeves on the oxidation of DCEE

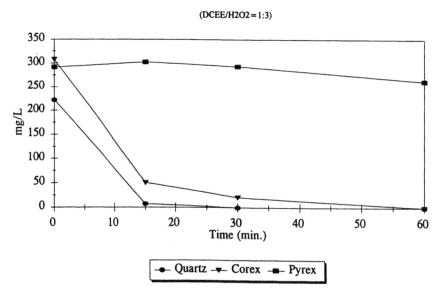

(DCEE/H2O2 = 1:3)

Figure 7. Effect of temperature on the oxidation of DCEE

(DCEE/H2O2=1:3.9)

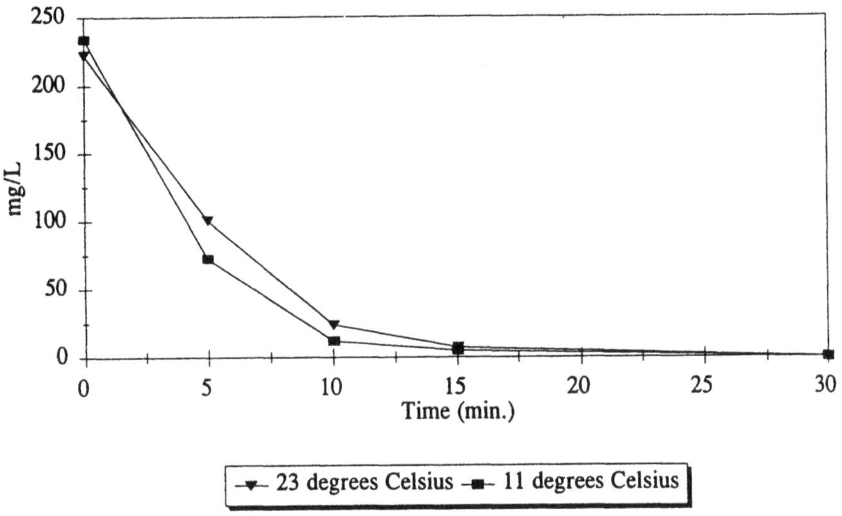

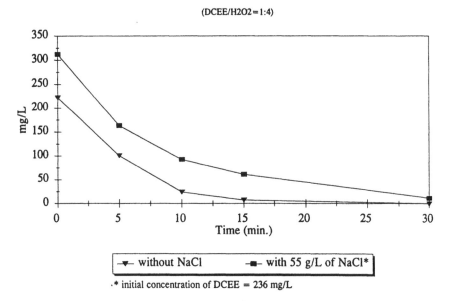

Figure 8. Effect of salinity on the oxidation of DCEE

(DCEE/H2O2=1:4)

.* initial concentration of DCEE = 236 mg/L

Figure 9. Effect of pH on the oxidation of DCEE

(DCEE/H2O2=1:3.9)

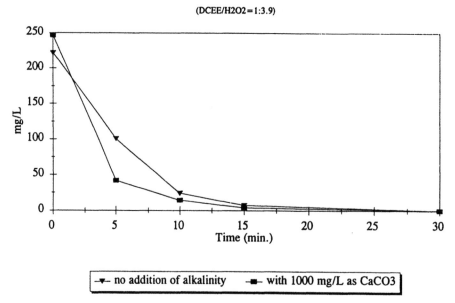

$-\blacktriangledown-$ pH=6.8/6.7 * $-\blacksquare-$ pH=6.7/2.8

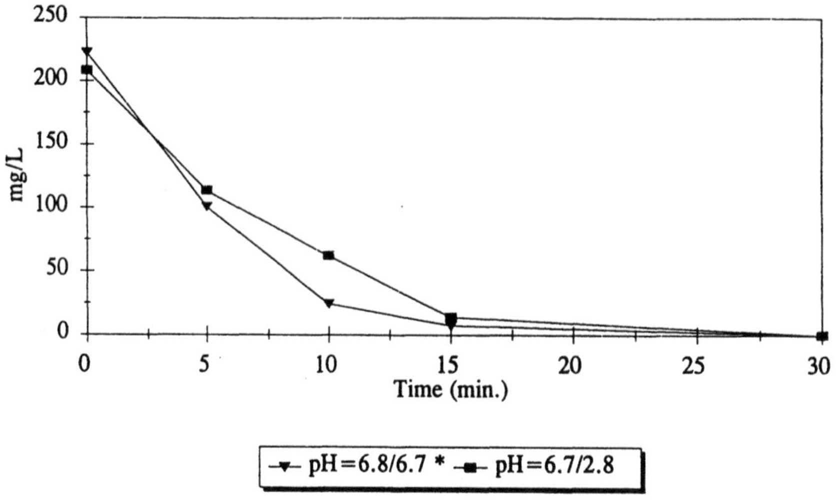

* pH before the reaction/pH after the reaction

Figure 10. Effects of alkalinity on the oxidation of DCEE

(DCEE/H2O2=1:3.9)

$-\blacktriangledown-$ no addition of alkalinity $-\blacksquare-$ with 1000 mg/L as CaCO3

126

OXIDATION OF CHLORINATED ETHANES

The concentrations of 1,2-DCA with respect to the reaction time under different DCA to H_2O_2 ratios is shown in Figure 11. The stoichiometric ratio of DCA to H_2O_2 is 1:5 (molar ratio) for the complete oxidation of DCA. Under this condition, over 99% of the 1,2-DCA at an initial concentration of 1230 ppm was oxidized after 60 minutes of irradiation. 1,1,2-TCA was the major oxidation intermediate with its concentration peaking out at 0.3 ppm (after 30 minutes of irradiation) and then decreasing to several ppb. Chloroethane, methylene chloride, 1,1,2,2-TetCA, and two isomers of tetrachlorobutane (1,2,3,4- and 1,2,2,4-) were also observed in the aqueous phase; however, their concentrations were always in the ppb levels during the course of reaction. At a lower H_2O_2 dose (DCA to H_2O_2 ratio of 1:2.04), the 60-minute removal efficiency of 1,2-DCA dropped to about 87%. The number and amount of oxidation by-products formed during the oxidation process also increased substantially. Therefore, the results indicated that a stoichiometric amount of H_2O_2 for the complete oxidation of DCA is required in order to prevent the formation of high molecular weight compounds which will tend to complicate the treatment process. It is also interesting to note that at the beginning of the reaction (about 5 minutes of irradiation), the oxidation rate of DCA at higher H_2O_2 doses was slower than that at lower H_2O_2 doses and the condition was reversed after 10 minutes of irradiation.

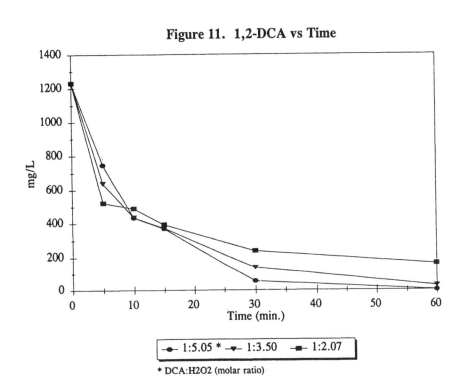

Figure 11. 1,2-DCA vs Time

The concentrations of 1,1,2-TCA with respect to the reaction time under different TCA to H_2O_2 ratios is shown in Figure 12. The stoichiometric ratio of TCA to H_2O_2 is 1:4 (molar ratio) for the complete oxidation of TCA. Under this condition, about 96.6% of the 1,1,2-TCA at an initial concentration of 370 ppm was oxidized after 60 minutes of

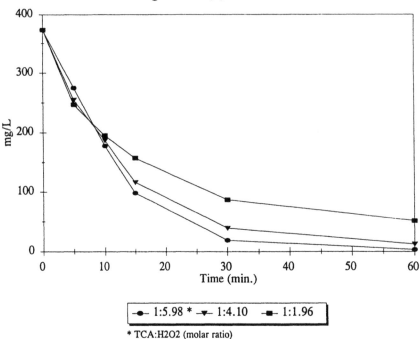

Figure 12. 1,1,2-TCA vs Time

Legend: 1:5.98 * — 1:4.10 — 1:1.96

* TCA:H2O2 (molar ratio)

irradiation. Methylene chloride was the major oxidation intermediate and its concentration was always at levels of ppb throughout the reaction. At a higher H_2O_2 dose, the oxidation rate increased only marginally with 99% removal of the initial TCA. The removal efficiency dropped to 86% at the ratio of TCA to H_2O_2 of 1:1.96 and numerous oxidation by-products, such as methylene chloride, tetrachloroethane, hexachlorobutene, and tetrachlropropene, were observed in the system.

The concentrations of 1,1,2,2-TetCA with respect to the reaction time under different TetCA to H_2O_2 ratios is shown in Figure 13. The stoichiometric ratio of TetCA to H_2O_2 is 1:3 (molar ratio) for the complete oxidation of TetCA. Under this condition, only about 79% of the TetCA at an initial concentration of 150 ppm was oxidized after 30 minutes of irradiation. Methylene chloride and chloroform were the major oxidation intermediates. At a higher H_2O_2 dose (TetCA:H_2O_2=1:5.14), the removal efficiency was improved to 89.5%. No high molecular weight by-products were formed even at TetCA:H_2O_2=1:2.54; however, numerous heavy oxidation by-products were observed in the system at TetCA:H_2O_2=1:1.21.

CONCLUSIONS

The UV peroxidation process provides an efficient removal of DCEE from the contaminated groundwater. At a higher H_2O_2 dose, over 99% of DCEE at an initial concentration of 220 mg/L can be destroyed within 30 minutes of irradiation with only minute amounts of oxidation by-products observed in the effluent. Volatile oxidation by-products were accumulating in the system when the H_2O_2 dose was low. Since no oxidation by-products of high molecular weights were observed even at very low H_2O_2 doses, it will be very cost-effective to combine the UV/peroxidation process with the existing air stripping unit for the treatment of DCEE. The residual amount of DCEE in the air-stripped

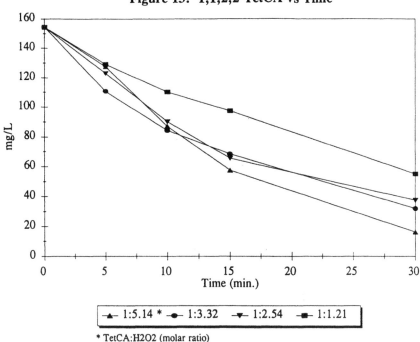

Figure 13. 1,1,2,2-TetCA vs Time

groundwater can be treated photo-chemically at low H_2O_2 doses. The generated oxidation by-products, which are more volatile than DCEE, can then be treated by recycling the groundwater back to the air stripping unit. Alkalinity and salinity at concentrations equivalent to those in the contaminated groundwater pose insignificant impact on the oxidation of DCEE. Temperature of the groundwater also shows negligible effects on the oxidation process when it goes down to 11 °C. The addition of buffer solutions to maintain the pH of the contaminated groundwater at about 7 is recommended since it may also be required by regulation before the treated groundwater being discharged into the receiving water body.

The results from the oxidation of 1,2-DCA, 1,1,2-TCA, and 1,1,2,2-TetCA indicated that these compounds are much more recalcitrant to the UV/peroxidation process than DCEE. Stoichiometic amounts of H_2O_2 for the complete oxidation of chloroethanes are required in order to prevent the formation of high molecular weight compounds except in the case of TetCA. Generally speaking, the UV/peroxidation process is not suitable as an alternative technology for the treatment of the contaminated groundwater because the high operating cost involved. For example, complete oxidation of 1,2-DCA will cost about $14/1000 gals of groundwater for the purchase of H_2O_2 alone (assuming $0.79/lb of H_2O_2, groundwater pumping rate of 30 gpm, and 8 working hours/day).

References

1. Liu, P.H., 1993. "Biodegradation of Selected Chlorinated Hydrocarbons in an Anaerobic Expanded-Bed Reactor (ANEBR)." Master's Thesis, Lamar University, Beaumont, TX.

2. Huang, F.Y.C., 1993. Unpublished Data.

3. McCoy and Associates, Inc., 1991. <u>The Clean Air Act Amendments of 1990: Statutory Requirements and Keyword Index</u>, McCoy and Associates, Inc.

4. Sundstrom, D.W., H.E. Klei, T.A. Nalette, D.J. Reidy, and B.A. Weir, 1986. "Destruction of Halogenated Aliphatics by Ultraviolet Catalyzed Oxidation with Hydrogen Peroxide." <u>Hazardous Waste and Hazardous Materials</u>, 3(1):101-110.

5. Froelich, E.M., 1991. "Advanced Chemical Oxidation of Contaminated Water Using the Perox-Pure™ Oxidation System." Proceedings of the First International Symposium on Chemical Oxidation: Technology for the Nineties, Nashville, TN.

6. Topudurti, K. and S. Matsui, 1992. "A UV/Oxidation Technology Demonstration to Treat Groundwater Contaminated with VOCs." <u>Water Sci. Technol.</u>, 25(11):347-354.

7. Sirabian, R., T. Sanford, and R. Barbour, 1992. "UV Peroxidation with Air Stripping for Optimized Removal of VOCs from Groundwater." Superfund '92: Proceedings of the National Conference, Washington, DC.

8. IT Corporation, 1991. "Cost-Effective, Alternative Treatment Technologies for Reducing the Concentrations of Methyl Tertiary Butyl Ether and Methanol in Groundwater." Final Report, Prepared for American Petroleum Institute, IT Corporation.

9. Bull, R.A. and J.D. Zeff, 1991. "Hydrogen Peroxide in Advanced Oxidation Processes for Treatment of Industrial Process and Contaminated Groundwater." Proceedings of the First International Symposium on Chemical Oxidation: Technology for the Nineties, Nashville, TN.

10. Hager, D.G., C.G. Loven, and C.L. Giggy, 1987. "Chemical Oxidation Destruction of Organic Contaminants in Groundwater." Superfund '87: Proceedings of the National Conference, Washington, DC.

11. Shimoda, S., H.W. Prengle, Jr., and J.M. Symons, 1993. "H_2O_2/Vis-UV Process for Treatment of Leachates, Contaminated Groundwater and Industrial Wastewater." Final Report Submitted to Gulf Coast Hazardous Substance Research Center, Beaumont, TX.

12. Topudurti, K.V., N.M. Lewis, and S.R. Hirsch, 1991. "UV/Oxidation Treatment of Contaminated Groundwater." Superfund '91: Proceedings of the National Conference, Washington, DC.

13. Levenspiel, O., 1989. <u>The Chemical Reactor Omnibook</u>, OSU Book Stores, Inc., Corvallis, Oregon.

K. M. HODGSON
T. R. LUNSFORD

Testing of Peroxidation Systems, Inc. perox-pure SSB-30

ABSTRACT

A facility is being designed and built at the U.S. Department of Energy (DOE) Hanford Site to treat water containing a variety of organic and inorganic compounds. An ultraviolet light/hydrogen peroxide system, manufactured by Peroxidation Systems, Inc. (PSI), has been chosen to destroy the organic compounds in the feed stream. The PSI perox-pure™* model SSB-30 has been tested by the Westinghouse Hanford Company (WHC) to provide data for permit documentation and to determine appropriate operating conditions. The destruction of the organic compounds was demonstrated with several feed compositions at different ultraviolet light exposures and hydrogen peroxide concentrations.

INTRODUCTION

In the past, the chemical processing facilities at the Hanford Site allowed large quantities of water with low levels of radionuclides to be discharged to shallow, sandy sediments below the ground surface. Favorable adsorption and filtration characteristics meant that most of the radionuclides were retained in a sediment column above the water table. Subsequently, the DOE implemented a policy requiring waste water treatment and minimization of radioactive and hazardous waste discharge. As a result, several projects have been initiated to treat major waste water streams and to remove radioactive and hazardous components from them.

One of the projects, the 200 Area Effluent Treatment Facility (ETF), will provide the facilities to treat and dispose of the 242-A Evaporator process condensate. Originally, ETF was designed to treat 242-A Evaporator process condensate and the process distillate discharge and the ammonia scrubber distillate of the Plutonium Uranium Extraction Plant (PUREX). Because the PUREX Plant was shut down by the DOE in January 1993, it

---

*perox-pure™ is a registered trademark of Peroxidation Systems, Inc.

Authors are K. M. Hodgson and T. R. Lunsford, Westinghouse Hanford Company, P.O. Box 1970, Richland, Washington, 99352.

is unlikely that the PUREX process distillate discharge and the ammonia scrubber distillate will be generated again. Therefore it is necessary only to address the treatment of 242-A Evaporator process condensate.

The process condensate is formed by the evaporation process which occurs when the 242-A Evaporator concentrates low-level waste stored in underground double-shell tanks. The double-shell tank waste and the process condensate are considered dangerous under the Washington Administrative Code (WAC) Chapter 173-303 [1] and the U.S. Environmental Protection Agency (EPA) regulations [40 Code of Federal Regulations 261] [2] because of the presence of the following:

- Halogenated spent solvents (F001 and F002) and nonhalogenated spent solvents (F003 and F005)
- Toxicity (WT02 per WAC 173-303)
- Radionuclides such as tritium, strontium-90, ruthenium-106, and cesium-137
- Inorganic compounds such as ammonia, potassium, silica, carbonate, chloride, and nitrate
- Organic compounds such as butyl alcohol, acetone, tetradecane, tridecane, and tributyl phosphate

The treatment system is illustrated in Figure 1. Treatment consists of the following steps:

1. Removal of suspended solids by filtration, and removal of organic compounds by the PSI ultraviolet light mediated oxidation process. The process utilizes hydrogen peroxide as an oxidant to promote the destruction of organic impurities.
2. Conversion of dissolved ammonia to ammonium sulfate by adding sulfuric acid to achieve a pH of 4 to 6.
3. Removal of most of the dissolved solids using a reverse osmosis unit. The retentate (concentrated stream) from this unit will be concentrated to the lowest possible volume.
4. Using an ion exchange/adsorption system for dissolved solids polishing. This will assure that the goals for removing radionuclides and dissolved solids will be met.
5. Adjusting the pH in-line from 6.5 to 8.5.

After being treated in the ETF, the effluent will be sent to holding tanks for sampling and analytical verification. If the effluent meets permit conditions, then it will be discharged to a State Approved Land Disposal Site.

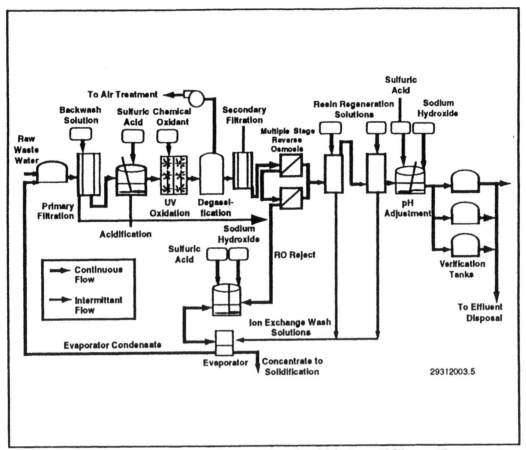

Figure 1. Treatment Flow Diagram for the 200 Area Effluent Treatment Plant.

Surrogate wastes were selected for testing because the 242-A Evaporator is shut down, and no waste water is available for pilot testing. Even when the evaporator restarts, the initial condensate produced will be too dilute to make it suitable for pilot testing.

Surrogate test solutions (STSs) were developed to contain carefully selected chemicals representing the types of constituents that will be of concern in the anticipated waste stream to be treated at the ETF. The solutions also contain various other organic and inorganic chemical groups expected to be found on the Hanford Site.

OBJECTIVE

The objective of testing was to demonstrate the concentration ranges of organic compounds that can be successfully treated by the ultraviolet light/hydrogen peroxide process in the ETF. The testing formed the basis of a petition to the EPA to delist the ETF effluent.

SCOPE

Testing included determining the destruction and removal efficiencies of the ultraviolet light/hydrogen peroxide, reverse osmosis, and ion exchange portions of the process. Only the results of the ultraviolet light portion of the testing are discussed. Testing was not intended to simulate a continuous ETF treatment train. Tests were performed in a semicontinuous mode where the surrogate test solution was sequentially processed through each step.

SURROGATE TEST SOLUTIONS

Surrogate Test Solutions composed of constituents characterizing the 242-A Evaporator process condensate were used for testing. Additional constituents were also included to expand the types of constituents that might be treated by the ETF. The STS constituents were selected from the 242-A Evaporator process condensate characterization data, the Hanford Site chemical inventory, and additional organic compounds representing various chemicals of regulatory concern. For a detailed explanation of STS development, see *C-018H Surrogate Test Solution* [3].

The STSs were tailored specifically to evaluate the ultraviolet oxidation rate of organic compounds, and the removal efficiency of inorganic compounds using reverse osmosis and ion exchange. Four solutions were developed and tested. For additional information, see the *Envelope Test Plan* [4] which is summarized as follows:

STS-1: STS-1 is considered "worst-case" waste. It consists of high concentrations of inorganic and organic constituents found in the 242-A Evaporator PC and identified in the Hanford chemical inventory [5].

STS-2: STS-2 is composed of typical concentrations of organic constituents found in the 242-A Evaporator process condensate and has high inorganic constituent concentrations. It also includes constituents identified in the Hanford Site chemical inventory [5]. This STS formulation was used primarily to evaluate the high inorganic concentration effect on the ultraviolet light mediated oxidation system.

STS-3: STS-3 is considered a more "typical case" waste. It consists of inorganic and organic constituents found in the 242-A Evaporator process condensate and identified in the Hanford Site chemical inventory [5].

STS-4: STS-4 is composed of high concentrations of inorganic constituents found in the 242-A Evaporator process condensate and identified in the Hanford Site chemical inventory [5] as well as organic constituents representing various chemical groups of regulatory concern. This STS formulation was used primarily to evaluate the high inorganic concentration effect on the organic oxidation of constituents of regulatory concern.

134

Each STS was prepared in a concentrated form and diluted to the correct volume before each test. Specific volatilization prevention techniques were implemented to minimize loss of organic compounds during STS preparation. The solutions were sampled immediately before the initiation of each test. The STS preparation is discussed in the following section.

SURROGATE TEST SOLUTION PREPARATION

Target constituent concentrations for each STS were reported in the *C-018H Surrogate Test Solution* document [3]. Hanford's Standards Laboratory in the 222-S Laboratory Complex prepared the concentrates for each solution based on the target constituent concentrations [3]. The prepared compositions of the STSs are given in Tables I and II. Each STS was prepared from concentrates of the inorganic, volatile organic, and semivolatile organic constituents and was then diluted to the full 200 gallon test volume.

Before beginning each test, the STS concentrate was fully diluted to 200 gallons. Exactly 195 gallons of deionized water was used as a diluent for the tests. The water was weighed in 55-gallon polyethylene containers on a calibrated scale. The mass was converted to volume using the density of water at the bulk temperature of the water. The water, except for approximately 5-gallons retained for rinsing purposes, was pumped into the ultraviolet light mediated oxidation feed/recycle tank.

TESTING

Tests were ordered as follows: STS-3, STS-1, STS-2, and STS-4. The STS-3 was chosen as the starting point because organic and inorganic concentrations are representative of the actual feed to the ETF. The STS-1 test, which followed, represented the most challenging feed with high concentrations of both organic and inorganic constituents. The STS-2 test had low organic and high inorganic concentrations. The STS-4 test was last because of its unique combination of organic compounds.

TEST CONDITIONS

The conditions of the ultraviolet light mediated oxidation testing are summarized in Table III. The oxidation time and hydrogen peroxide concentrations selected were based on the organic and inorganic compositions of the STSs. Oxidation time and hydrogen peroxide concentration were adjusted to compensate for variations in the feed composition and to demonstrate the flexibility and capability of the ultraviolet light mediated oxidation process to treat feed streams with varying organic and inorganic

TABLE I. PREPARED COMPOSITIONS OF THE SURROGATE TEST SOLUTIONS (ORGANIC).

Analyte	STS-1 Prepared Concentration (ppb)	STS-2 Prepared Concentration (ppb)	STS-3 Prepared Concentration (ppb)	STS-4 Prepared Concentration (ppb)
Acetone	24,983	2,498	2,498	
Acetonitrile				2,498
Acrolein				2,497
Aniline				4,994
Benzene	2,495	250	249	
Bis (2-chloroethyl) ether				2,491
Bis (2-ethylhexyl) phthalate				100
n-Butanol	100,353	9,990	10,035	14,092
Carbon Tetrachloride	2,498	251	251	
Chloroform	2,504	250	250	
1,4-dichlorobenzene				4,994
gamma-BHC				1,998
Hexachloroethane				2,498
Methyl ethyl ketone	4,997	500	494	
Methyl isobutyl ketone	4,996	500	496	
Naphthalene	2,498	250	250	
Nitrobenzene				4,991
N-nitroso-di-n-propylamine				2,497
Pentachlorophenol				2,497
Phenol	2,496	250	250	
Pyridine	2,499	251	252	
Toluene	2,498	249	251	
Tributyl phosphate	14,985	9,989	10,007	7,056
1,1,1-Trichloroethane	2,507	251	249	
Tridecane	999	999	989	999

TABLE II. PREPARED COMPOSITIONS OF THE SURROGATE TEST
SOLUTIONS (INORGANIC).

Analyte	STS-1 Prepared Concentration (ppb)	STS-2 Prepared Concentration (ppb)	STS-3 Prepared Concentration (ppb)	STS-4 Prepared Concentration (ppb)
Aluminum	5,020	4,995	1,249	4,955
Ammonium	2,509,075	2,497,490	49,968	2,497,464
Arsenic	2,510	2,497	407	2,497
Barium	201	200	100	200
Beryllium	100	100	75	100
Cadmium	1,004	999	500	999
Cesium	502	500	100	499
Chromium	2,510	2,497	499	2,498
Copper	1,004	999	489	999
Iron	502	499	250	499
Lead	100	100	50	100
Mercury	100	100	50	100
Nickel	1,004	999	499	999
Ruthenium	764	761	152	761
Selenium	1,005	985	858	1,000
Silicon	4,415	4,392	2,196	4,391
Silver	201	200	100	200
Sodium	44,314	44,086	12,230	44,091
Strontium	502	499	100	499
Vanadium	502	499	100	499
Zinc	1,004	999	499	999
Carbonate	10,046	10,006	2,498	9,990
Chloride	1,760	1,752	635	1,752
Cyanide	2,008	1,999	499	1,998
Fluoride	25,099	24,975	4,995	24,975
Nitrate	1,058,532	1,057,874	61,364	1,057,814
Sulfate	6,282,532	6,250,222	101,080	6,250,221

TABLE III. ULTRAVIOLET LIGHT MEDIATED OXIDATION TESTING PARAMETERS
AND SAMPLE TIMES.

Parameter	STS-1	STS-2	STS-3	STS-4
Initial hydrogen peroxide concentration (mg/L)	500	250	250	500
Hydrogen peroxide concentration (mg/L) after 1/3 of oxidation time	50 after 204 minutes	50 after 111 minutes	50 after 44 minutes	200 after 22 minutes
Oxidation time	46	25	10	5
Test duration (minutes)	613	333	133	67
Protocol characterization sampling times	Every 10 minutes of oxidation time	Every 7 minutes of oxidation time	Every 3 minutes of oxidation time	Every 1.5 minutes of oxidation time
Protocol characterization sampling times in terms of test duration (minutes)	0, 133, 267, 400, 533, and 613	0, 93, 187, 280, and 333	0, 40, 80, and 133	0, 20, 40, and 67

compositions. The pH and turbidity of the feed solutions were monitored throughout the tests, and the temperature of the feed solution was maintained at 22 °C. The feed flow rate for each test was 40 gallons per minute.

TEST APPARATUS

The pilot test unit is the Peroxidation Systems, Inc. **perox-pure** Model SSB-30. The unit is constructed of stainless steel and is equipped with six, high intensity, 5-kilowatt ultraviolet lamps. Each lamp is individually enclosed in a quartz sheath and each is wired to separate a switch so that any one can be independently operated, depending on the desired ultraviolet light energy input. Figure 2 shows the flow diagram with the process instrumentation for the ultraviolet light mediated oxidation pilot-scale testing.

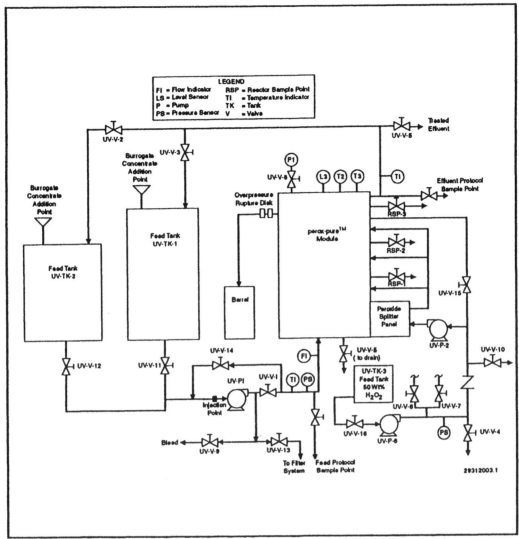

Figure 2.   Ultraviolet Light Oxidation Test Apparatus Flow Diagram

The pilot-scale ultraviolet light mediated oxidation unit does not have a quartz sheath cleaning mechanism. The quartz sheath can be cleaned chemically or by removing the sheath from the reactor and wiping it with an absorbent cloth. Before and after each test run, the quartz sheaths were cleaned to maintain them as close as possible to the original condition for each test.

A hydrogen peroxide flow splitter can be used to inject hydrogen peroxide into the pilot-scale reactor at three locations. A pump meters the correct amount of hydrogen peroxide to the flow splitter, and the flow splitter divides preset amounts of hydrogen peroxide into the first, middle, and last sections of the reactor. For these tests the hydrogen peroxide was equally divided to each reactor section.

SAMPLE ANALYSIS

With one exception, the analytical methods used for organic analysis followed EPA SW-846 methods [6]. The semivolatile organic compound analysis was performed using the Contract Laboratory Program (CLP) analysis method [7]. This method is similar to SW-846 Method 8270. The CLP analysis method was preferred for semivolatile organic compounds because quinones and resorcinol compounds were lost in the SW-846 Method 8270 extraction process.

Throughout the tests, hydrogen peroxide levels were determined with colormetric indicator strips which compared the color from a strip dipped into the sample with the color of a strip dipped into a standard of known concentration. Standards of 50, 70, 90, and 100 milligrams per liter (mg/L) were obtained for the Hanford Standards Laboratory. The range of the strips is 0 to 100 mg/L. Samples, in which the hydrogen peroxide concentrations exceeded the concentrations of the standards, were diluted to bring the samples within the range of the standards as verified by the indicator strips.

RESULTS

Tables IV through VII summarize the organic data collected from these tests, and Figure 3 shows the results for total organic carbon in which the concentration is plotted versus time. Plotted concentrations and sample times were normalized to the concentration at time zero and the length of the test respectively so that destruction of the organic compounds could be compared between tests. The oxidation times and hydrogen peroxide concentrations were chosen to compensate for the type and concentration of the inorganic constituents and the type and concentration of the organic compounds incorporated into the STSs. By normalizing the concentrations and oxidation times, the capability of the ultraviolet light mediated oxidation system to compensate for the different feed composition of the STSs can be seen. Figure 3 shows that by varying oxidation times and hydrogen peroxide concentrations, the ultraviolet light mediated oxidation system was able to decrease the total organic carbon concentration more than 80 percent before 40 percent of the testing time had expired for STS-1, STS-2, and STS-3. These three surrogates had the same organic compounds but in different concentrations or in different inorganic matrices. Although the fourth surrogate behaved differently, the end result was the same.

Figures 4 and 5 summarize the destruction achieved. Figure 4 compares the amount of each organic chemical destroyed in STS-1, STS-2, and STS-3. The organic and inorganic compounds were the same for the first three surrogates, but the concentrations of the chemical compounds varied between the surrogate solutions. Figure 5 shows the destruction of the organic compounds in STS-4.

TABLE IV. ORGANIC ANALYTICAL DATA FOR STS-1.

Analytes	Target concentration	System feed	Sample time = 133[b]	Sample time = 267[b]	Sample time = 400[b]	Sample time = 533[b]	Sample time = 613[b]
	ppb	ppb	ppb	ppb	ppb	ppb	ppb
Acetone	25,000	14,000	3,200	150	26	10[a]	10[a]
Benzene	2,500	1,700	2.5	3	1	5[a]	1
n-Butanol	100,000	120,000	100[a]	100[a]	100[a]	100[a]	100[a]
Carbon tetrachloride	2,500	480	37	13	4	2	2
Chloroform	2,500	1,900	875	290	120	52	29
Methyl ethyl ketone	5,000	5,300	10[a]	10[a]	10[a]	10[a]	10[a]
Methyl isobutyl ketone	5,000	5,800	10[a]	10[a]	10[a]	10[a]	10[a]
Toluene	2,500	1,000	1	5[a]	5[a]	5[a]	5[a]
1,1,1-Trichloroethane	2,500	1,300	725	220	86	32	16
Naphthalene	2,500	1,900	10[a]	10[a]	10[a]	10[a]	10[a]
Phenol	2,500	2,700	10[a]	10[a]	10[a]	10[a]	10[a]
Pyridine	2,500	100[a]	50[a]	50[a]	50[a]	50[a]	50[a]
Tributyl phosphate	15,000	15,000	20[a]	20[a]	20[a]	20[a]	20[a]
Tridecane	1,000	780	430	150	85	40	21
TOC	107,830[c]	90,890	26,225	6,400	224	894	679

[a]Analyte reported below detection limit, detection limit reported.

[b]Sample times reported in minutes of test duration, oxidation time is equal to test duration time multiplied by (15/200).

[c]Target TOC is the sum of the carbon in the organic compounds.

TABLE V. ORGANIC ANALYTICAL DATA FOR STS-2.

Analytes	Target concentration	System feed	Sample time = 93[b]	Sample time = 187[b]	Sample time = 280[b]	Sample time = 333[b]
	ppb	ppb	ppb	ppb	ppb	ppb
Acetone	2,500	3,900	650	120	50	34
Benzene	250	210	2	1	1	5[a]
n-Butanol	10,000	36,000	100[a]	100[a]	100[a]	100[a]
Carbon tetrachloride	250	120	59.5	22	12	9
Chloroform	250	260	130	59	34	25
Methyl ethyl ketone	500	820	10[a]	10[a]	10[a]	10[a]
Methyl isobutyl ketone	500	470	10[a]	10[a]	10[a]	10[a]
Naphthalene	250	160	10[a]	10[a]	10[a]	10[a]
Toluene	250	150	5[a]	5[a]	5[a]	5[a]
1,1,1-Trichloroethane	250	170	115	44	25	16
Phenol	250	210	10[a]	10[a]	10[a]	10[a]
Pyridine	250	100[a]	50[a]	50[a]	50[a]	50[a]
Tributyl phosphate	10,000	8,000	20[a]	20[a]	20[a]	20[a]
Tridecane	1,000	530	315	150	93	72
TOC	16,144[c]	11,680	2,890	1,355	788	234

[a]Analyte reported below detection limit, detection limit reported.
[b]Sample times reported in minutes of test duration, oxidation time is equal to test duration time multiplied by (15/200).
[c]Target TOC is the sum of the carbon in the organic compounds.

TABLE VI. ORGANIC ANALYTICAL DATA FOR STS-3.

Analytes	Target concentration	System feed	Sample time = 40[b]	Sample time = 80[b]	Sample time = 133[b]
	ppb	ppb	ppb	ppb	ppb
Acetone	2,500	1,800	200	10[a]	10[a]
Benzene	250	270	8	4	3
n-Butanol	10,000	7,100	100[a]	100[a]	100[a]
Carbon tetrachloride	250	150	67.5	41	19
Chloroform	250	290	104.5	31	6
Methyl ethyl ketone	500	78	10[a]	10[a]	10[a]
Methyl isobutyl ketone	500	390	10[a]	10[a]	10[a]
Toluene	250	180	4	5[a]	5[a]
1,1,1-Trichloroethane	250	240	100	28	5[a]
Naphthalene	250	130	10[a]	10[a]	10[a]
Phenol	250	180	10[a]	10[a]	10[a]
Pyridine	250	50[a]	50[a]	50[a]	50[a]
Tributyl phosphate	10,000	4,900	20[a]	20[a]	20[a]
Tridecane	1,000	130	195	77	150
TOC	16,144[c]	10,500	1,440	280	180

[a]Analyte reported below detection limit, detection limit reported.

[b]Sample times reported in minutes of test duration, oxidation time is equal to test duration time multiplied by (15/200).

[c]Target TOC is the sum of the carbon in the organic compounds.

143

TABLE VII. ORGANIC ANALYTICAL DATA FOR STS-4.

Analytes	Target concentration	System feed	Sample time = 20[b]	Sample time = 40[b]	Sample time = 67[b]
	ppb	ppb	ppb	ppb	ppb
Acetone	0	18	250	11	140
Acetonitrile	2,500	12	14.5	14	13
Acrolein	2,500	2,400	190	21	20[a]
Benzene	0	16	16.5	12	13
n-Butanol	10,000	8,900	650	100[a]	100[a]
Chloroform	0	8	7	7	5
Tetrachloroethylene	2,500	1,200	465	340	240
Tetrahydrofuran	5,000	5,300	210	5[a]	5[a]
1,1,2-Trichloroethane	2,500	2,400	2,100	1,900	1,000
Aniline	5,000	2,700	355	14	20[a]
Bis(2-chloroethyl) ether	2,500	1,700	270	12	10[a]
Bis (2-ethylhexyl) phthalate	100	59	41	17	14
1,4-dichlorobenzene	5,000	1,900	35.5	5[a]	10[a]
gamma-BHC	2,000	1,400	1,300	670	190
Hexachloroethane	2,500	930	855	710	570
Nitrobenzene	5,000	3,300	145	2	10[a]
N-nitroso-di-n-propylamine	2,500	1,450	5	10[a]	10[a]
Pentachlorophenol	2,500	1,500	6	20[a]	20[a]
Tributyl phosphate	10,000	4,800	63	20[a]	20[a]
Tridecane	1,000	360	360	140	140
TOC	32,918[c]	21,000	17,4450	13,170	792

[a]Analyte reported below detection limit, detection limit reported.
[b]Sample times reported in minutes of test duration, oxidation time is equal to test duration time multiplied by (15/200).
[c]Target TOC is the sum of the carbon in the organic compounds.

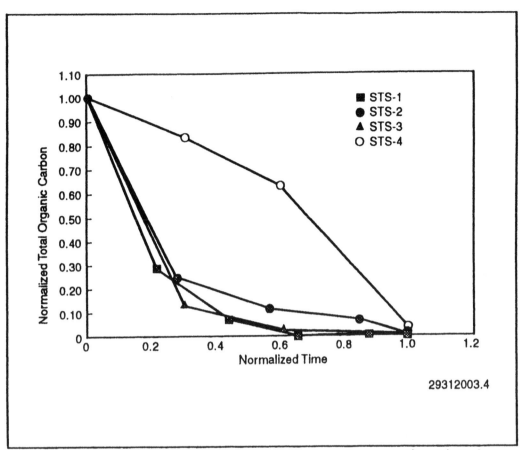

Figure 3. Normalized Total Organic Carbon Concentration Plotted
versus Normalized Time.

The results of each test are discussed below in the order in which the tests were
conducted:

STS-1

Most of the organic compounds except the chlorocarbons and tridecane, were
reduced to less than their detection limits. More than 99 percent of all compounds were
destroyed.

The analytical results for tridecane seem to be most consistent in this test. Possibly,
the substantial amount of 1-butanol added (100,000 micrograms per liter) helped dissolve
the tridecane in the water. Therefore, consistent amounts of tridecane were available in
the water for extraction and injection into the analytical instrument.

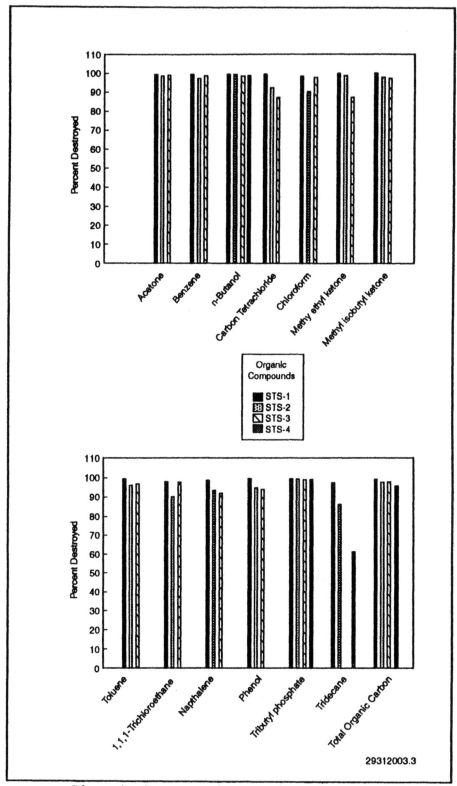

Figure 4. Summary of Destruction Efficiencies.

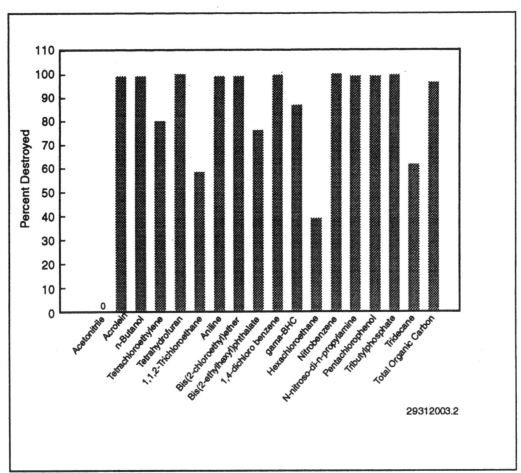

Figure 5. Destruction Efficiencies for STS-4.

STS-2

Most of the organic compounds were reduced to less than detection limits. Exceptions included the following: acetone, carbon tetrachloride, chloroform, 1,1,1-trichloroethane, and tridecane.

STS-3

All the organic compounds except the chlorocarbons and tridecane were reduced to less than their detection limits. The tridecane results were so erratic that conclusions cannot be drawn. More than 87 percent of each chlorocarbon was destroyed. The chlorocarbons and straight chain hydrocarbons appeared to be the most difficult to destroy.

STS-4

The chlorocarbons and tridecane were the most difficult to destroy. Thirty-nine percent of the hexachloroethane was removed. Of the compounds that are not straight chain hydrocarbons or chlorocarbons, only bis (2-ethyl hexyl) phthalate had less than 99 percent of the original concentration destroyed.

Acetone and chloroform, which may have been produced as degradation products, were subsequently destroyed. Chloroform may have been introduced as a contaminant in the deionized water used to prepare the solutions.

CONCLUSIONS

This testing showed that by adjusting operating conditions the **perox-pure** ultraviolet light mediated oxidation system was capable of processing feed streams that deviated significantly from the design basis. By varying the oxidation time and hydrogen peroxide concentration it was possible to obtain high levels of destruction for a wide variety of organic compounds in a matrix of low levels and high levels of inorganic compounds.

The chlorocarbon and straight chain hydrocarbon compounds were the most difficult to destroy, but significant levels of destruction were achieved.

ACKNOWLEDGMENTS

The authors would like to acknowledge the contributions of Tim P. Moberg, Craig V. King, Don E. Gana, Deirell D. Paine, and Phillip M. Olson.

REFERENCES

1. WAC 173-303, "Dangerous Waste Regulations," Washington State Department of Ecology, Olympia, Washington.

2. 40 CFR 261, "Identification and Listing of Hazardous Waste," Title 40, *Code of Federal Regulations*, Part 261, U.S. Environmental Protection Agency, Washington, D.C.

3. WHC, 1992, *C-018H Surrogate Test Solution*, WHC-SD-C018H-TS-001, Rev. 0, Westinghouse Hanford Company, Richland, Washington.

4. WHC, 1993, *Envelope Test Plan*, WHC-SD-C018H-TP-009, Rev. 0, Westinghouse Hanford Company, Richland, Washington.

5. WHC, 1990, *Inventory of Chemicals Used at Hanford Site Production Plants and Support Operations (1994-1980)*, WHC-EP-0172, Revision 1, Westinghouse Hanford Company, Richland, Washington.

6. EPA, 1987, *Test Methods for Evaluating Solid Waste, Physical/Chemical Methods*, SW-846, Update II issued 1990, U.S. Environmental Protection Agency, Washington, D.C.

7. EPA, 1991, *Statement of Work for Organic Analysis Contract Laboratory Program* EPA OLM 01.8, U.S. Environmental Protection Agency, Washington, D.C.

S. LAURA YEH
CARMEN LEBRON

Photochemical Oxidation for the Remediation of Ordnance Contaminated Groundwater

ABSTRACT

The Naval Facilities Engineering Service Center (NFESC), formerly the Naval Civil Engineering Laboratory (NCEL), has recently completed a treatability study to investigate the applicability of advanced oxidation processes (AOP), primarily photochemical oxidation technology, for the treatment of ordnance contaminated groundwater. The following was accomplished during this treatability study: a laboratory bench-scale test, a vendor bench-scale test, and an on-site pilot study. The feasibility study demonstrated that AOP could successfully degrade the ordnance compounds TNT and RDX to below ppb levels. The adjustment of pH was the only pretreatment used. The ultimate objective was to compare full-scale AOP costs of to a similar sized GAC system.

During the laboratory study, UV/H_2O_2, UV/O_3, UV photolysis, and O_3/H_2O_2 were studied for effectiveness in removing TNT and RDX from contaminated groundwater samples, with focus on reaction chemistry. During the vendor study, UV/O_3 and UV/H_2O_2 were tested by three AOP vendors using their bench-scale equipment. UV/O_3 treated the ordnance compounds most efficiently, requiring the least energy and shortest treatment times, and was recommended for further investigation in a pilot study. The pilot study, conducted at a Navy site in the State of Washington, allowed for further optimization of parameters (e.g. UV intensity, oxidant dose, UV/oxidant ratios, and pH) investigated during the bench-scale vendor study which decreased full-scale treatment cost estimates obtained during bench-scale testing. However, further system optimization is needed to apply the photochemical oxidation technology cost-competitively with GAC for the particular site investigated in this study.

INTRODUCTION

NFESC has been investigating the use of photochemical oxidation technologies for removing organic compounds in groundwater for the past several years. The Navy owns many sites where groundwater contamination by organic solvents, petroleum hydrocarbons, and ordnance compounds exists. The photochemical oxidation technology

S. Laura Yeh and Carmen Lebron, Naval Civil Engineering Laboratory, Port Hueneme, California 93041, USA

represents an innovative ex-situ treatment method which may be applied to the cleanup of these sites in conjunction with a pump and treat remedial strategy. In contrast to GAC and air-stripping technologies, photochemical oxidation technologies remove organic compounds via contaminant destruction rather than concentration on another phase which requires further treatment. In implementing the photochemical oxidation technology to a site, treatability testing is necessary for several reasons: to determine whether the compounds of concern may be successfully degraded; to ensure that during treatment any by-products which are formed are removed; and to optimize system design and operating costs for this energy-intensive treatment process.

In photochemical oxidation processes, UV light and an oxidant (typically ozone (O_3) or hydrogen peroxide (H_2O_2)) are employed to create a highly reactive environment for compound degradation. There are three pathways by which degradation can occur: UV photolysis, direct reaction with O_3 and/or H_2O_2, and the synergistic combination of UV/oxidant. Hydroxyl radicals formed from UV/oxidant interactions can attack compounds which are typically recalcitrant to oxidation.

PROJECT DESCRIPTION & APPROACH

The AOP treatability study was conducted for an NPL site, Site F, at the Naval Submarine Base Bangor, Washington. Groundwater contamination at Site F has resulted from the disposal of ordnance-contaminated water from demilitarization operations into an unlined lagoon. This site is well-suited for the use of photochemical oxidation since it possesses low turbidity and low metals concentrations. TNT (2,4,6-Trinitrotoluene) and RDX (Hexahydro-1,3,5 - trinitro-1,3,5-triazine) were the principal contaminants that were investigated for treatment effectiveness. HMX (Octahydro-1,3,5,7-Tetranitro-1,3,5,7-Tetrazocine) was also present at low levels in the water, however, it was not identified as a compound requiring treatment and therefore not investigated in this project.

The goal of the Site F AOP project was to develop an AOP system that would be cost-competitive with GAC treatment for the site under investigation by optimizing process parameters. The experimental work for the project was performed by outside agencies and vendors. The project was conducted from September 1991 to June 1993, in three phases. In the first phase, which consisted of a laboratory study performed by the Illinois State Water Survey, the goal was simply to establish the treatability of the ordnance contaminated groundwater using different advanced oxidation processes and to study the reaction chemistry so that vendors could optimize their systems in the next phase of the project. In the second phase, three vendors were selected to conduct bench-scale tests using actual groundwater samples to see if their systems could meet treatment goals (see below). In the third phase, one vendor system was selected for further optimization during an on-site pilot test.

Prior to this study, no treatment standards existed for the compounds of concern. The following Washington state Model Toxic Control Act (MTCA) standards were established for this study: 2.9 ppb for TNT, 0.8 ppb for RDX, and 1 ppb for TNB (by product of TNT degradation).

GROUNDWATER CHARACTERISTICS

In each work phase, the untreated groundwater samples were analyzed by ISWS as well as other commercial laboratories. A complete summary of groundwater characteristics obtained for Phase I work is shown in Table I below. The variation in ordnance concentrations between the three work phases is approximately ± 0.5 ppm, and is shown in Table II. The ordnance concentrations in this table reflect averages from multiple determinations. Site F groundwater contains low levels of ordnance, at 6 ppm

TABLE I - ISWS CHARACTERIZATION OF UNTREATED GROUNDWATER SAMPLES FOR PHASE I WORK

Organic Constituents	
Component	*Concentration*
2,4,6-Trinitrotoluene (TNT)	6040 μg/l
Hexahydro-1,3,5 - trinitro-1,3,5-triazine (RDX)	648 μg/l
1,3,5-Trinitrobenzene (TNB)	389 μg/l
Total Organic Carbon (TOC)	5580 μg/l
Inorganic Species	
Species	*Concentration*
Iron	0.048 mg/l
Manganese	0.144 mg/l
Calcium	8.99 mg/l
Magnesium	4.57 mg/l
Sodium	7.18 mg/l
Aluminum	0.02 mg/l
Barium	0.018 mg/l
Chromium	0.008 mg/l
Potassium	1.91 mg/l
Strontium	0.114 mg/l
Silicon	14.9 mg/l
Ammonium	41.9 mg/l
Fluoride	< 0.1 mg/l
Chloride	1.8 mg/l
Nitrate (as nitrate)	143 mg/l
Sulfate	18.9 mg/l
Alkalinity (as CaCO$_3$)	54.4 mg/l
Inorganic Carbon	20.8 mg/l

153

TABLE II - ISWS CHARACTERIZATION OF ORDNANCE COMPOUNDS IN GROUNDWATER SAMPLES

Study Phase: Well Sampling Date:	III Feb 1993	II March 1992	I Dec 1991
Ordnance Compound	Concentration in $\mu g/l$		
RDX	457	563	648
TNB	355	422	389
TNT	6520	6470	6040

and 0.6 ppm of TNT and RDX respectively. Metals, which are oxidized during AOP treatment and can cause lamp fouling, are present in Site F groundwater at concentrations below 1 ppm (iron and manganese). Alkalinity and ammonium ion concentrations, at 54.4 and 41.9 mg/l respectively, pose a concern due to potential hydroxyl radical scavenging, i.e. bicarbonate and ammonia are more reactive towards hydroxyl radical than ordnance compounds, thus wasting hydroxyl radical. The high nitrate levels are also a concern because they are in excess of drinking water standards, and nitrate concentrations could potentially increase from AOP reactions.

RESULTS & DISCUSSION

PHASE I

During the laboratory study, ISWS performed twelve batch experiments (or semi-batch since oxidant usually fed continuously) using an 8.5-L stirred tank photochemical reactor (STPR) which could support a maximum of four-5.3 watt (low intensity, 254 nm output) UV lamps. One experiment was conducted using a 1 kilowatt (high intensity, short wavelength) lamp for which a special 5 liter photochemical reactor was constructed. The number and type of UV lamp(s) used and dosage of oxidant(s) was varied with experimental runs. In addition, pH adjustment and sparging at the adjusted pH were employed to remove hydroxyl radical scavengers (bicarbonate and ammonia) from solution was employed in most runs. Nine runs examined various UV and oxidant combinations and include a UV photolysis experiment and two ozone/peroxide experiments; three runs examined the behavior of nitrate and nitrite in the reaction system. Since the focus was reaction chemistry, experiments were designed to optimize data collection rather than to achieve the most rapid destruction of contaminants. A summary of the nine UV/oxidant runs (Runs 1-4 and 8-12) that were performed is provided in Table III. Sparging to remove bicarbonate was performed for Runs 3, 8, 9, 10, 11 and 12.

Analyses were performed using high performance liquid chromatography (HPLC). A modified EPA Method 8330 was used as the HPLC protocol for separation and quantitation of ordnance compounds. Preconcentration methods were developed during the laboratory study to meet the MTCA standards, as the maximum quantification limits

154

TABLE III - EXPERIMENTAL CONDITIONS FOR PHASE I RUNS

Run No.	AOP Configuration	Run Length (min)	Initial pH	UV dose (watt/L)	Applied Ozone Dose (mg/l-min)	H₂O₂ Dose (mg/l-min)
1	UV Photolysis	60	6.9 (unadjusted)	2.2	0	0
2	Ozone/Peroxide	90	9.9	0	1.4	0.20
3	UV/Ozone	90	4.2	1.25	1.4	0
4	UV/Peroxide	90	6.8 (unadjusted)	2.2	0	340 mg/l (initial dose)
8	Ozone/Peroxide w/ Bicarbonate sparging	90	9.0 (pH 4.0 during 30-min. sparging)	0	1.4	0.20
9	UV/Ozone	90	4.2	1.25	2.1	0
10	UV/Peroxide	180	4.0	2.2	0	10 mg/l (initial dose)
11	UV/Peroxide using 1 kW UV lamp	45	4.1	10 (approximate)	0	340 mg/l (initial dose)
12	UV/Ozone	120	4.2	0.63	1.4	0

afforded using the modified EPA protocol were 20 ppb and 50 ppb for TNT and RDX respectively. Total Organic Carbon (TOC) was also measured as a parameter of treatment efficiency for the removal of unidentified by-products from TNT and RDX degradation. Although an extensive investigation of by-products was beyond the scope of this study, one persistent compound, TNB, was detected as one of the last TOC components to be degraded. The experimental results will be discussed in terms of destruction of parent ordnance compounds, by-products formed from parent compound degradation, and ozone mass transfer.

TNT and RDX Destruction Effectiveness

TNT and RDX were rapidly destroyed by any AOP which employs UV photolysis, including UV photolysis alone. Destruction of >99% of the TNT in 30 minutes at a UV dose of 1.25 watts/L was typical. Removal of RDX was even more rapid. Ozone/peroxide treatment was unsatisfactory for the removal of TNT and RDX. This was thought to be due to scavenging of OH radicals by ammonia at the elevated pH (9.8) at which ozone/peroxide was applied. TNT and RDX removals for each UV/oxidant experimental run are shown in Figures I and II.

TOC/TNB Destruction Effectiveness

UV/ozone was found to be the fastest and most efficient of the AOPs for removal of oxidation by-products, measured as TOC. Of the three UV/ozone experiments conducted (Runs 3, 9 and 12), Run 12, with the lowest UV/ozone ratio, was the most efficient in achieving TOC removal, where TOC removal efficiency was measured as the moles of TOC removed per mole of ozone used.

UV/peroxide treatment using low peroxide concentrations was found to be very slow for TOC removal when the low-intensity 254 nm lamps were used. Constant peroxide addition to maintain the low steady-state concentration while irradiating with a high-intensity short-wavelength lamp showed some promise, but was not optimized during this laboratory study. Little or no TOC removal was achieved using UV photolysis alone. The TOC removals achieved with each experimental run is shown in Figure III.

Of all the AOPs that were investigated, only UV/ozone treatment resulted in the complete or near-complete destruction of TNB. The behavior of TNB during AOP treatment as TNT is degraded is complicated by the fact that it is initially present in the groundwater samples at concentrations of approximately 389 ppb or 0.4 ppm. TNB evolution curves during AOP treatment are shown in Figure IV.

Mass Transfer

Mass transfer efficiency was measured as a ratio of utilized ozone dose to applied ozone dose. Ozone mass transfer efficiencies were measured for all runs where ozone was utilized, i.e. Runs 3, 9, and 12 corresponding to UV/ozone, and Runs 2 and 8,

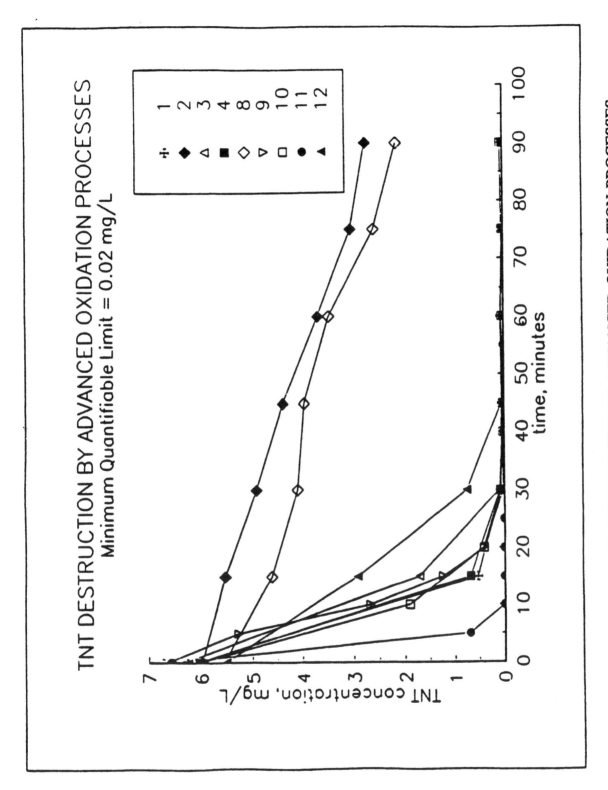

FIGURE I - TNT DESTRUCTION BY ADVANCED OXIDATION PROCESSES

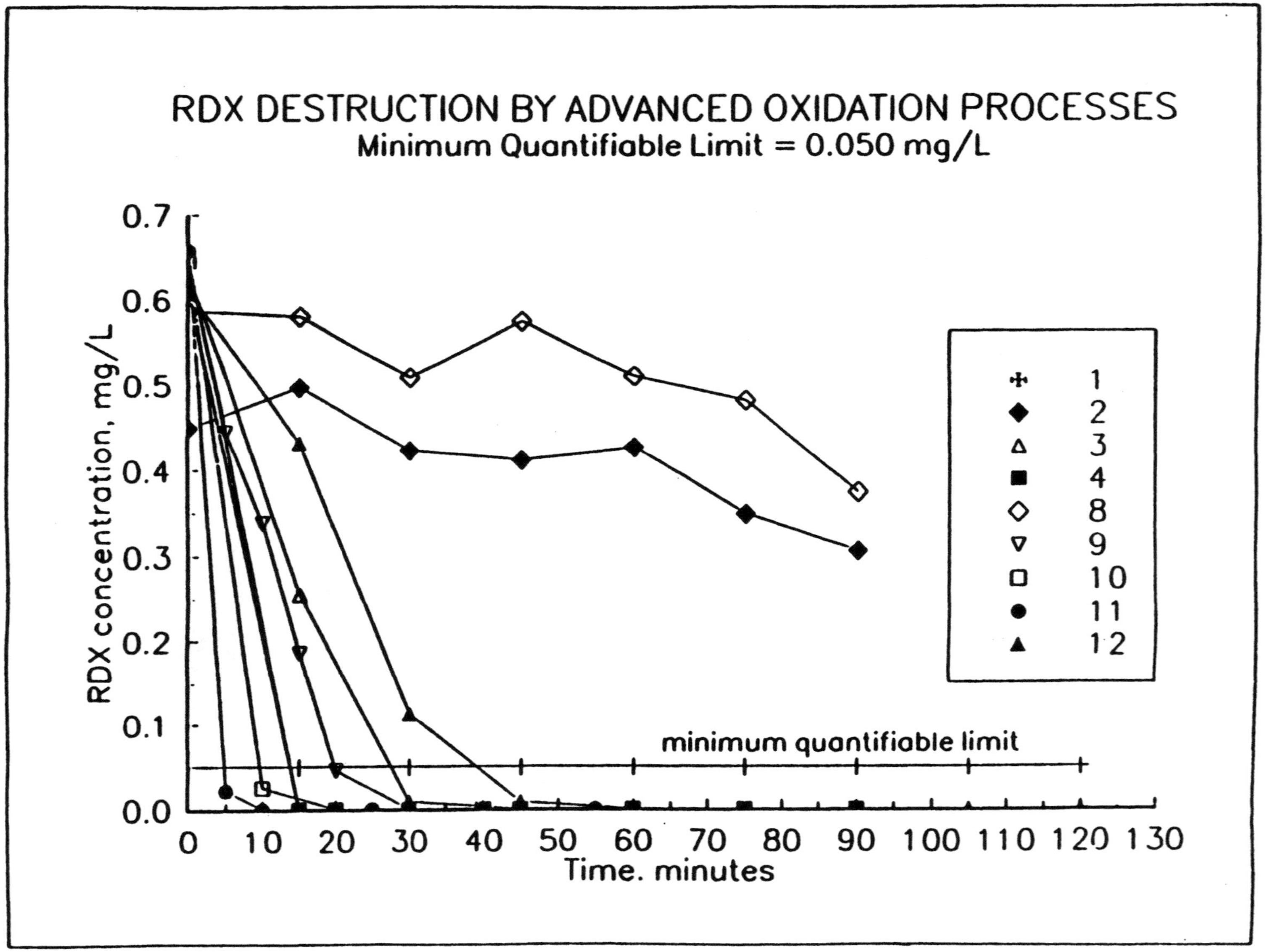

FIGURE II - RDX DESTRUCTION BY ADVANCED OXIDATION PROCESSES

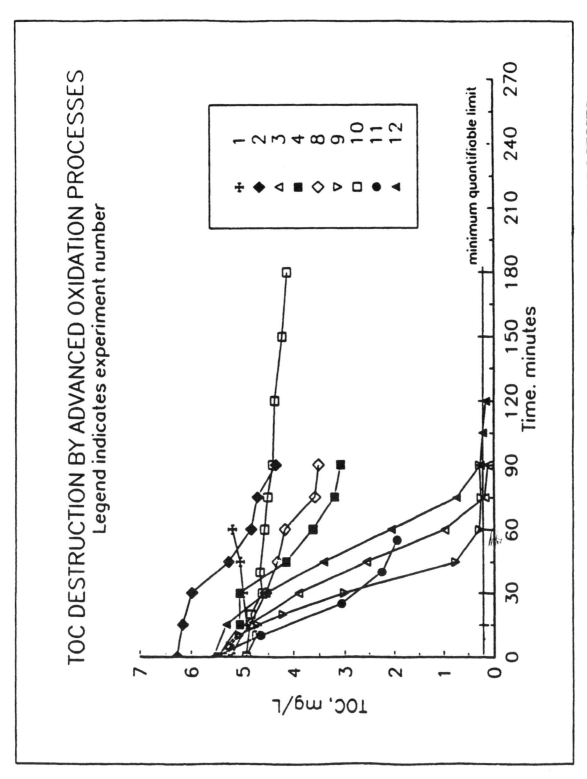

FIGURE III - TOC DESTRUCTION BY ADVANCED OXIDATION PROCESSES

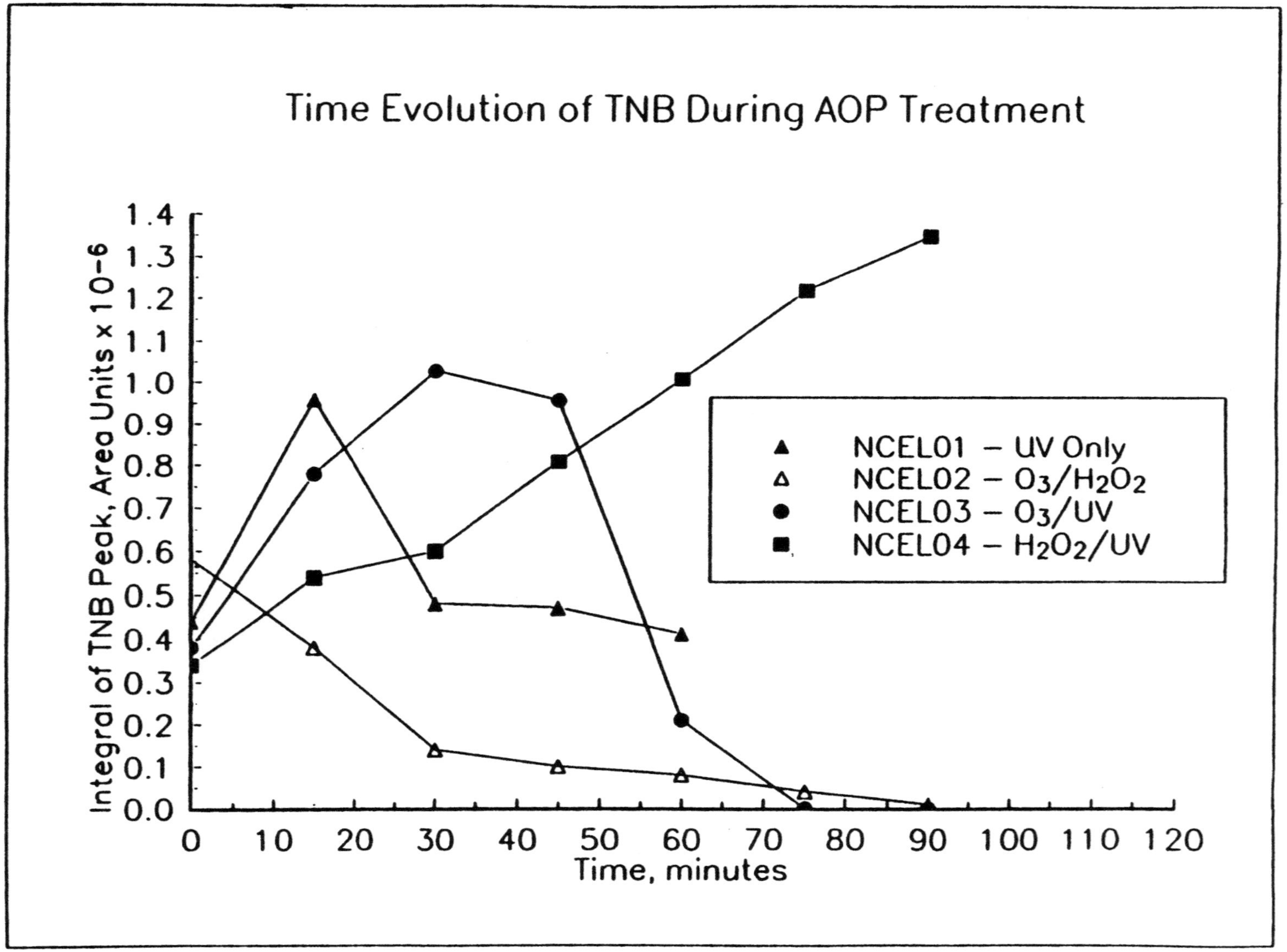

Time Evolution of TNB During AOP Treatment

FIGURE IV - TNB BEHAVIOR DURING VARIOUS AOP TREATMENTS

160

corresponding to ozone/peroxide. Runs 3 and 9 showed the highest ozone mass transfer efficiencies, at 90+ %. Runs 2 and 8 showed mass transfer efficiencies of approximately 60%. In Run 12, the ozone mass transfer efficiency began at around 60% and then increased to 85%. Since the TOC destruction efficiency was about 30% higher in Run 12 although mass transfer efficiency was about 30% lower initially, it is hypothesized that the additional ozone that was being transferred in Runs 3 and 9 was being photolyzed to peroxide and being used in unproductive radical-scavenging reactions. Ozone mass transfer efficiencies are shown in Figure V.

Removal of Nitrate and Ammonia

Neither nitrate nor ammonia were removed by AOP treatment. Therefore additional treatment is necessary to remove these compounds from Site F groundwater.

PHASE II

The purpose of the Phase II vendor studies was to determine the treatment capabilities of commercial AOP units, and to identify reaction parameters that were effective in achieving destruction of the target (RDX, TNT, and TNB) contaminants. Three vendors with established UV/oxidation technologies and high flow rate commercial units currently in operation (Peroxidation Systems, Inc., Ultrox International, and Solarchem Environmental Systems) were selected to conduct bench-scale AOP experiments. Each vendor was allotted a limited number of samples for ordnance analyses that they were allowed to apportion between experimental runs as needed. These samples were analyzed by ISWS with a 5- to 7- day sample turnaround so that vendors could use the results to plan their subsequent tests. The vendors used their most successful runs to provide preliminary full-scale treatment costs.

Each vendor used a cylindrical reactor with an axially mounted UV lamp to perform the experiments. Both Peroxidation Systems and Solarchem used water recirculation to promote mixing in the reactor and to maintain a constant reactor volume while experimental samples were being collected. Ultrox used a magnetic stirrer for mixing instead of water recirculation. Peroxidation Systems used three different lamps with different UV intensities in its experiments. Solarchem used a single lamp but reduced the lamp output in several runs by using a stainless steel shade. Ultrox used the same unshielded UV lamp in all of their experimental runs.

A summary of vendor bench-scale equipment specifications and operating parameters is provided in Table IV. Some information concerning the individual test units, particularly the UV lamps, was considered proprietary and not provided. Each vendor performed six to ten experimental runs on the Site F groundwater samples. Peroxidation Systems tested UV/peroxide in each of its eight experiments, although ozone and catalyst were also added in separate experiments. The pH was adjusted in five runs. Solarchem tested UV/ozone treatment in four of their six experiments. Catalyst addition was performed on one of the UV/ozone runs. UV/peroxide and a proprietary catalytic ozonation (no UV) process were tested in the other two runs. Solarchem did not perform any pH adjustment. Ultrox tested UV/ozone treatment with pH adjustment in

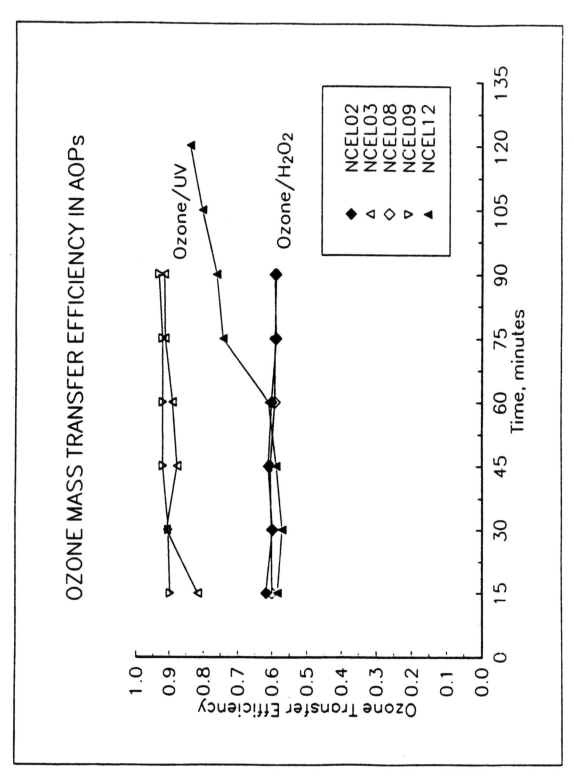

FIGURE V - OZONE MASS TRANSFER EFFICIENCY FOR SELECTED RUNS

162

TABLE IV - SUMMARY OF VENDOR BENCH TEST EQUIPMENT SPECIFICATIONS AND EXPERIMENTAL PROCEDURES

	Peroxidation Systems	Solarchem	Ultrox
Equipment Parameter			
UV lamp type	Normal, Medium-High and High intensity, medium pressure	High intensity, short wavelength	Low pressure, mercury arc
Lamp UV output in watts	proprietary	1,000 (see note)	14
Reactor type	Horizontal cylindrical	Vertical cylindrical	Vertical cylindrical
Reactor volume in liters	not provided	6	2.4
Experimental Procedures			
Approximate water volume charged in liters	6	24	2
Water circulation rate in gpm	4	10-20	not applicable
Typical experimental run length in min	2	25-300	40

Note: Solarchem partially shielded its 1kW lamp to reduce UV output for some experimental runs

each of its runs. Hydrogen peroxide was also added in one experiment, but catalyst addition was not tested. The most successful vendor runs will be discussed. Destruction curves are not provided because the number of samples collected between vendors per experimental run varies significantly, so comparable curves can not be shown. Vendors typically monitored reaction progress by an easily measured parameter in their own in-house lab such as TOC and performed other analyses in-house or at local laboratories.

Experimental Results

The experimental conditions for the most successful bench-scale runs from each vendor is shown in Table V. These runs came the closest to meeting and/or exceeding treatment goals. Where used, catalyst addition was found not to improve contaminant destruction. TNB was the parameter that presented the most difficulty in terms of meeting treatment goals. Peroxidation Systems was able to achieve the destruction of TNB using the high intensity UV lamp, high H2O2 doses and pH adjustment. However,

163

TABLE V - RUN CONDITIONS FOR MOST SUCCESSFUL VENDOR BENCH-SCALE EXPERIMENTS

Vendor	UV/oxidant configuration	Run Length (min)	Initial pH	UV dose (watts/l)	Ozone Dose (mg/l-min)	H2O2 Conc. (mg/l)
Peroxidation Systems	UV/H2O2	3	4.0	note (1)	0	100 mg/l (initial dose)
Solarchem	UV/ozone	25	5.5	18.5 note (2)	20.5	0
Ultrox	UV/ozone	45	4.0	9.5	1.6	0

Note (1): Not provided, Peroxidation Systems used a high intensity UV lamp in this run.
Note (2): This is the lamp UV output divided by total water volume used, 25 liters, in the experiment. If the reactor volume is used, with approximately 5.4 liters of water in the 6 liter reactor, this becomes 85 watts/l.

the energy costs required for a UV/H2O2 system that was capable of meeting treatment goals eliminated this configuration as an alternative for further investigation.

Solarchem showed that by decreasing the UV to ozone ratio, the destruction of ordnance compounds was improved. In comparing Solarchem and Ultrox process parameters in their most successful experiments, it is evident that Ultrox was able to use substantially less ozone and UV energy than Solarchem to achieve treatment goals. The UV dose used by Solarchem is twice that used by Ultrox, and if Solarchem's UV dose is examined for the reactor volume alone, instead of for the full 25 liter batch volume, Solarchem's UV dose becomes eight times higher than Ultrox. There is a similar disparity in ozone doses.

The tremendous difference in the UV and ozone doses that were successfully used by Ultrox and Solarchem to achieve the destruction of ordnance compounds reflects differences in approach for AOP treatment. Ultrox commercial units use low intensity UV lamps (up to 65 watts/lamp) in large volume reactors (thousands of gallons) whereas Solarchem systems use staged reactors, with each reactor (< 50 gallons, estimated) containing a single high intensity UV lamp (15-30 kW/lamp). In examining preliminary full-scale treatment costs, each vendor was found to have its own standard method for scaling-up from bench-scale to full-scale units. Solarchem used a direct scale-up of UV and ozone dosages, since its full-scale system is geometrically similar to its bench-test unit, i.e. a single lamp per reactor, while Ultrox used a more empirical approach based on past experience, due to significant geometric differences between its bench and full-scale unit. Although significant discussion could be engaged on the pros and cons of these two vendor systems and scale-up methods, it is beyond the scope of this paper.

PHASE III

The goal of the on-site pilot study was to test and evaluate AOP treatment and equipment performance on a larger scale, to perform further optimization of treatment parameters, and to obtain treatment costs for a full-scale 200 and 400 gpm AOP system. As in Phase II, the focus was on determining ultimate treatment efficiencies under different reaction conditions for scale-up purposes rather than collecting data with which to study reaction chemistry. During this pilot study, Solarchem performed 14 batch (or semi-continuous) runs using a 16-liter recirculating batch reactor which could support a single, axially mounted, 1 or 4 kW UV lamp. The 4 kW lamp extended the full length of the reactor whereas the 1kW lamp extended only 31 percent of the reactor length and was mounted in the middle of the reactor. The lamp was protected by a cylindrical quartz sleeve. In addition, during UV lamp operation, an automatic mechanical cleaning device passed over the quartz sleeve every 15 minutes to remove any accumulated deposits. During the tests, water was recirculated between a 600 gallon process tank and the reactor at a rate of 8 to 22 gpm. Typical batch volumes were 150 or 300 gallons and run lengths were 1 to 3 hours.

Two rounds of tests were conducted, so that there would be flexibility in designing the final experimental runs based on the analytical results obtained from the first round of experiments. An on-site HPLC (Round 1) and UV spectrophotometer (Rounds 1 and 2) were used to measure the destruction effectiveness of the various runs being conducted. Round 1 consisted of four preliminary runs (P-1 through P-4) and six formal runs (SES-7 through SES-12). During the preliminary runs, equipment functioning, potential synergistic effect of oxygen in accelerating contaminant destruction, the necessity of UV for contaminant destruction, and the need for ozone bubbles in the reactor was tested. In both Phases I and II, the UV to ozone ratio was identified as a process parameter that affected treatment effectiveness, particularly for the removal of TNB. The trend was that decreasing the UV to ozone ratio increased the effectiveness of TNB removal; in addition, Phase I showed that the effectiveness of ozone mass transfer decreases as the UV to ozone ratio is decreased. Thus, in the first round of experimental runs, three UV/ozone ratios were tested and will be referred to as high, medium and low. Although Solarchem had not adjusted pH during its Phase II experiments, the results of other Phase I/II experimenters had suggested that lowering pH may be beneficial. Solarchem also hypothesized that oxygen-fed ozone could provide treatment advantages over air-fed ozone. Therefore, other parameters also investigated during the first round of experiments were initial pH and oxygen versus air feed for ozone production.

In the final round, four experimental runs (SES-13 to SES-16) were conducted. The first run (SES-13) was performed to repeat a successful Round 1 run. Two runs, SES-14 and SES-16, were performed to examine the effect of increasing gas and liquid residence times. To examine the effect of increasing gas residence time a medium UV/ozone ratio was used with a 1 kW lamp and the lowest ozone dosage possible as determined by the process equipment. To examine the effect of increasing liquid residence time, a low UV/ozone ratio run was performed with the minimum possible water recirculation rate. In addition, an experimental run to examine ozonation at high pH (SES-15) was attempted as an effort to eliminate the costs of UV from the full-scale AOP system.

A summary of the experimental run conditions used during the pilot test are shown in Table VI. Unless otherwise specified, all runs were conducted using oxygen-fed ozone. Samples were collected from sample ports located up- and downstream of the reactor typically at the beginning and end of each run and at regular intervals in between. This

TABLE VI - EXPERIMENTAL CONDITIONS FOR PILOT STUDY

Run ID	Description	Batch Water Volume in gallons	Initial pH	Water Recirc. Rate in gpm	UV Lamp Power in kW	Ozone Mass Rate in lb/hr	UV/O3 ratio (see note)	Run Length in hr
Preliminary Runs:								
P-1	Test equipment	300	6.5	19	4.1	1.65	2.5	2.0
P-2	UV w/O2 injection	300	6.5	19	4.1	na	na	1.1
P-3	Ozonation at neutral pH	300	6.5	19	0.0	1.68	0.0	1.0
P-4	UV w/dissolved O3	300	6.5	22	4.1	1.60	2.6	1.3
Round 1 Runs:								
SES-7	High UV/O3; neutral pH	300	6.5	20	4.1	0.43	9.5	2.8
SES-8	High UV/O3; low pH	300	4.5	20	4.1	0.39	10.4	2.8
SES-9	Medium UV/O3; neutral pH	300	6.5	19	4.1	1.65	2.5	2.7
SES-10	Low UV/O3; neutral pH	150	6.5	18	1.2	1.71	0.7	4.0
SES-11	SES-7 w/ Air-fed O3 generation	300	6.5	20	4.1	0.38	10.8	2.6
SES-12	Low UV/O3; low pH	150	4.5	18	1.2	1.42	0.8	3.0
Round 2 Runs:								
SES-13	Confirmation of SES-12	150	4.5	17	1.0	1.30	0.8	0.8
SES-14	Med UV/O3, Low gas flow rate	150	4.5	17	1.0	0.43	2.3	3.0
SES-15	Ozonation at high pH	300	12	17	0.0	1.54	na	2.0
SES-16	Low UV/O3, low water recirc rate	150	4.5	8	1.0	1.42	0.7	1.0

Note: UV/O3 ratio is defined as the number of kWh of electrical energy consumed by the UV lamp power supply per pound of ozone fed to the reactor.

166

allowed the vendor to estimate a per-pass conversion of contaminant for use in scale-up calculations.

Experimental Results

A summary of the preliminary runs (P-1 to P-4) is given as follows: P-1 showed that the equipment functioned properly; P-2 showed that oxygen had no effect on the rate of photolysis; P-3 that direct ozonation without UV light was ineffective; and P-4 that ozone bubbles were required in the reactor, rather than being dissolved in the batch tank.

Round 1 results confirmed the previous observation that lowering the UV to ozone ratio improved the speed and efficiency of ordnance compound removal, but decreased ozone mass transfer efficiency. Neither of the high UV/ozone ratio runs conducted at the unadjusted pH (SES-7 and SES-11) achieved TNB destruction to the required 1 ppb concentration. When the pH was adjusted for the high UV/ozone ratio run in SES-8, treatment goals were met, thus reaffirming that pH adjustment enhanced treatment efficiency. Figure VI shows the effect of pH adjustment on treatment. Figure VII shows TNB concentration vs. time for high, medium, and low UV/ozone ratios. With the exception of SES-7 and SES-11, all Round 1 runs met treatment goals. SES-12, which used the lowest possible UV to ozone ratio, (0.7 - 0.8 kWh/lb), the 1 kW lamp, and pH adjustment, was identified as the most optimum Round 1 run. UV/ozone ratios lower than that used in SES-12 could not be attempted due to limitations imposed by the equipment in varying UV/ozone doses. In addition, UV and ozone doses at the UV to ozone ratio used in SES-12 could not be decreased beyond the limitations of the 1 kW lamp. Another consideration was that as UV to ozone ratio was decreased the ozone mass transfer efficiency became too low to be cost-effective in a full-scale system.

Round 2 results showed the repeatability of SES-12 in the success of SES-13. SES-15, high pH ozonation, could not adequately remove AOP by-products. SES-14 and SES-16 were performed in an attempt to determine the effects of increasing liquid and gas residence times to improve ozone mass transfer, however equipment limitations made it difficult in quantify the individual effects of these parameters. Since ozone is introduced with an in-line static mixer in Solarchem's pilot system, decreasing liquid flow rates and gas flow rates changes the ozone mass transfer efficiency in the system. In SES-14, which has the same liquid recirculation rates but a lower gas flow rate than SES-12, mass transfer efficiency is expected to increase, simply from hydrodynamic considerations which would cause a smaller ozone bubble size in the reactor. Thus, it is not clear whether the increase in ozone mass transfer efficiency in SES-14 compared to SES-12 can be attributed to a longer gas residence time.

In SES 16, with lower liquid recirculation rates and the same gas flow rates as SES-12, the ozone mass transfer efficiency would have been expected to decrease, as this combination would create a larger ozone bubble size in the reactor. In addition, by increasing the liquid residence time, the potential for phase separation of the ozone and water increases which would tend to reduce the residence time of the gas in the reactor which would tend to make ozone mass transfer less efficient. Thus, despite the increased liquid residence time, the results for SES-16 show very poor ozone mass transfer efficiency. Table VII compares the ozone mass transfer efficiencies and other

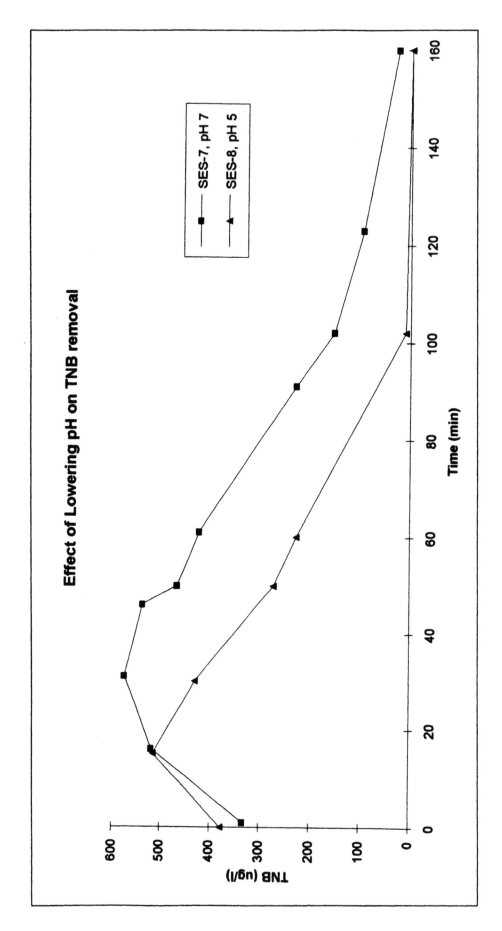

FIGURE VI - EFFECT OF LOWERING pH on TNB REMOVAL

168

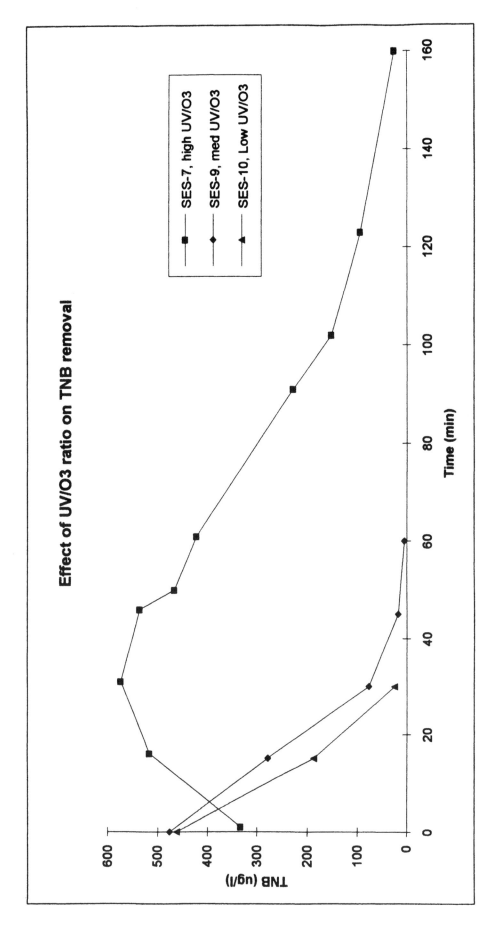

FIGURE VII - EFFECT OF UV/OZONE RATIO ON TNB REMOVAL

TABLE VII - COMPARISON OF SES-12, SES-14, AND SES-16

Run ID	Description	Initial pH	UV/O3 ratio (1)	Treatment Time in min (2)	Ozone capture eff. (%)	Reactor UV density in W/L (3)	Per-Pass Residence Time in sec
SES-12	Low UV/O3, low pH	4.5	0.8	35	35	75	3.5
SES-14	Low UV/O3, low pH, low gas flow rate	4.5	2.3	69	44	63	8.2
SES-16	Low UV/O3, low pH, low water flow rate	4.5	0.7	35	14 (4)	63	4.8

(1) Measured as kWh of lamp power output divided by lbs of ozone fed to the reactor
(2) Estimated time to reach treatment goals.
(3) Watts of UV lamp power to reactor volume in liters.
(4) The accuracy of this number is not as high as the other ozone capture efficiencies measured due to the error associated with measuring gas-phase ozone concentration, and the relatively small difference between influent and effluent conentrations.

run parameters for SES-12, SES-14, SES-16. SES-13 is not shown since it is almost identical to SES-12. The ozone mass transfer efficiency for SES-16 could not be accurately measured since the inlet and outlet ozone concentrations in the gas stream were very close to each other.

As noted in Phase II, there is a very large difference between the UV and ozone doses used in Solarchem's system compared to Ultrox's system. UV and ozone doses used in Phase III are also well in excess of those used in Phase I. At such high UV and ozone doses, ozone mass transfer efficiencies during Phase III were almost half (45% compared to 90%) of what was obtained in Phase I. Although there is something to be said about the difference in data collection goals, and differences in methods of ozone measurement employed during Phases I and III, the disparity is great enough that further reduction of UV and ozone doses should be examined.

FULL SCALE TREATMENT COSTS

Full-scale O&M costs were halved from original estimates obtained using Phase II testing results. Although this made AOP O&M costs competitive with GAC, capital costs were still not competitive with GAC for Site F groundwater. Vendor estimates for treating Site F groundwater using a 200 gpm AOP system were $1.3M at the influent concentrations tested during the treatability study compared to $200,000 for a comparable 200 gpm GAC system. Costs for a 400 gpm treatment systems were roughly double that of 200 gpm systems for both AOP and GAC.

CONCLUSIONS & RECOMMENDATIONS

The AOP treatability study demonstrated that photochemical oxidation processes can successfully degrade ordnance compounds in Site F groundwater. Both TNT and RDX are readily removed, while TNB, which is produced from TNT degradation, is a persistent compound that should be studied for further optimization of treatment parameters. During testing, UV/ozone proved to be the most efficient AOP for removing TNB. Low UV to ozone ratios were observed to increase treatment efficiency. Although further decreases in UV to ozone ratios were discouraged during pilot testing, due to the extremely low ozone mass transfer efficiencies being obtained in Solarchem's high UV intensity pilot system, using lower UV and ozone doses may allow this ratio to be further decreased to the benefit of treatment efficiency. From comparison of Phase III to Phase I and II results, further reduction of UV and ozone doses from that tested in Phase III is possible and should be examined. Additional study of liquid and gas residence times and reactor UV density on the removal of ordnance compounds and ozone mass transfer is recommended.

REFERENCES

1. Naval Civil Engineering Laboratory, 1992a. "Final Batch Testing Report, Deliverable No. P20." USN NCEL Managed, Bremerton, Washington, Pollution Abatement RDT&E Support Task, GSA Task number POC171028, ADP Support Services - Scientific/Engineering, GSA Contract Number GS09K-90-BHD0001. Dated April 21, 1992.

2. Naval Civil Engineering Laboratory, 1992b. "Final Phase II Report, Deliverable No. P38." USN NCEL Managed, Bremerton, Washington, Pollution Abatement RDT&E Support Task, GSA Task number POC171028, ADP Support Services - Scientific/Engineering, GSA Contract Number GS09K-90-BHD0001. Dated July 28, 1992.

3. Naval Civil Engineering Laboratory, 1993a. "Final Vendor Report for On-Site, Pilot-Scale for AOP Treatment Testing, Deliverable No. D7." USN NCEL Managed, Bremerton, Washington, Pollution Abatement RDT&E Support Task, GSA Task number POC172028, ADP Support Services - Scientific/Engineering, GSA Contract Number GS09K-90-BHD0001. Dated April 30, 1993.

4. Naval Civil Engineering Laboratory, 1993b. "Final Phase III Report, Deliverable No. D10." USN NCEL Managed, Bremerton, Washington, Pollution Abatement RDT&E Support Task, GSA Task number POC172028, ADP Support Services - Scientific/Engineering, GSA Contract Number GS09K-90-BHD0001. Dated May 21, 1993.

5. Naval Civil Engineering Laboratory, 1993c. "AOP Technology Package: A Comparison of AOP and GAC Treatment of Low-Level Ordnance Contamination in Site F Groundwater, Deliverable No. D12." USN NCEL Managed, Bremerton, Washington, Pollution Abatement RDT&E Support Task, GSA Task number POC172028, ADP Support Services - Scientific/Engineering, GSA Contract Number GS09K-90-BHD0001. Dated June 28, 1993.

DANIEL C. SCHMELLING
KIMBERLY A. GRAY

Photocatalytic Transformations of TNT in Titania Slurries: An Analysis of the Role of Interfacial Nitrogen Reduction Utilizing γ-Radiolysis

ABSTRACT

The photodegradation of TNT in a TiO_2 slurry reactor was studied as both a potential treatment technique for the remediation of water with munitions contamination and to gain further insight into the behavior of a nitroaromatic compound in a photocatalytic system. Photocatalytic and direct photolytic reactions were compared by evaluating rates and extent of TNT transformation and mineralization in the presence and absence of oxygen. Nitrate, nitrite, and ammonium ion concentrations were determined and mass balances on carbon and nitrogen were performed for the catalytic system. The gamma radiolysis of photolyzed and non-photolyzed solutions of TNT was conducted to elucidate the extent of photoreduction occurring in TNT photolysis. TNT is photolytically labile and was transformed rapidly under each set of photochemical conditions. Transformation by-products were destroyed readily under oxygenated catalytic conditions using near UV radiation (λ >340 nm) with greater than 90% mineralization achieved in 120 minutes. Additionally, significant amounts of ammonium ion were produced under these conditions which suggests that the photocatalytic transformation of TNT involves both oxidative and reductive steps. In contrast, little by-product destruction occurred under direct photolytic conditions. The γ-radiolysis data indicate that although direct photoreduction of the aromatic nitrogen occurs, most of the ammonium ion observed during TNT photocatalysis occurs as a result of interfacial reduction at the catalyst surface.

INTRODUCTION

TNT (2,4,6-trinitrotoluene) is a conventional secondary explosive which has been widely used by the military since the beginning of this century. Extensive contamination of ground water, sediments, and soils by TNT has occurred due to past practices for disposal of the high volume waste streams generated by munitions plants. Such contamination is of concern because TNT is toxic to many organisms including humans, is a known mutagen [1] and is listed as a priority pollutant by the U.S. Environmental Protection Agency [2]. More than 1200 sites with explosives contamination have been identified by the U.S. Department of Defense to require potential cleanup and more than 87% of these have groundwater contamination [3].

1. D.C. Schmelling and K.A. Gray, Dept. of Civil Engineering and Geological Sciences, University of Notre Dame, Notre Dame, IN 46556

173

The remediation of TNT contamination continues to be the subject of much research. In general, the parent compound TNT is relatively labile both biochemically and photochemically whereas the reaction by-products have proven more difficult to degrade. Activated carbon is often used to treat process waters at munitions plants as well as to remediate explosives contaminated groundwater [3,4]. However, this technology is non-destructive and may be expensive to operate [3]. In addition, ultimate disposal of the TNT and spent carbon usually involves incineration [5]. Incineration is among the most effective ways to destroy TNT, especially on soils but poor public approval and concern over air emissions have motivated the search for alternative treatment techniques.

During the past decade there has been significant growth in the utilization of semiconductor particulate systems as photocatalysts for carrying out the photochemical transformation and mineralization of organic compounds. These transformations are facilitated when charge separations are induced in a large band-gap semiconductor by excitation with ultra-bandgap light. The electron-hole pair so formed can then migrate to the catalyst particle surface and participate in interfacial redox reactions. Moreover, degradation has been achieved by TiO_2 photocatalysis using near UV and solar light as opposed to the high energy UV required in photochemical processes such as UV-peroxide/ozone [6,7]. This feature in combination with the fact that complete mineralization is often achieved suggests that TiO_2 photocatalysis may offer a number of advantages for the treatment of TNT contaminated water.

All of the photocatalytic experiments presented in this work were performed utilizing near UV light ($\lambda > 340$ nm). The purpose of blocking radiation below 340 nm was to eliminate as much direct photolysis of TNT as possible while still transmitting light sufficiently energetic to induce charge separation in TiO_2. Moreover, by attenuating higher energy light the reaction conditions were restricted to the range of UV energies predominant in solar radiation. These results, therefore, will allow consideration of the efficacy of utilizing solar light to photocatalytically degrade TNT.

In this paper a preliminary analysis of the photodegradation of TNT in a TiO_2 slurry reactor is presented. Photocatalysis is evaluated in comparison to direct photolysis by examining rates and extent of TNT transformation and mineralization in both aerated and deaerated environments. Consideration is given to the organic and inorganic products of these photoreactions and the inorganic nitrogen species are quantified and evaluated. Gamma-radiolysis has been employed to examine photoreduction occurring in the photolysis of TNT. The purpose of this work was to evaluate TiO_2 photocatalysis as a process for the remediation of TNT contaminated water and also to initiate a more detailed study of the photocatalytic behavior of nitroaromatic compounds.

EXPERIMENTAL METHODS

Standard grade 2,4,6-trinitrotoluene was purchased from Chemservice. Titanium dioxide (TiO_2) was obtained from DeGussa (P25). The annular reactor used for this work consisted of a jacketed Pyrex glass finger inserted into a Pyrex glass cylinder and is fully described elsewhere [8]. The UV radiation source, a 450 W mercury lamp (Conrad-Hanovia), was inserted into the center of the Pyrex glass finger. A filter solution of 0.25 M $CuSO_4$ and 0.25 g/L 2,7-dimethyl-3,6-diazacyclohepta-1,6-diene perchlorate was circulated through the Pyrex finger jacket to eliminate radiation with $\lambda < 340$ nm [see ref. 9 for transmission spectra] and to provide cooling.

A standard experiment involved the addition of 50 mg of dried TNT to 1 L of ultrapure

(18 M $\Omega \cdot$cm) water and is described in more detail elsewhere [8]. Experiments were also conducted under a N_2 atmosphere and with no TiO_2. Additional experiments were run with $NaNO_3$ or $NaNO_2$ and *tert*-butyl alcohol and with NH_4Cl instead of TNT.

Gamma radiolysis was performed with a Shepherd 109 irradiator. This had a 10,000 curie ^{60}Co source and the dose rate as determined by Fricke dosimetry was 6.1 kgray/hr. All irradiated samples were sparged with N_2O prior to γ-radiolysis.

TNT concentration and some by-product information were acquired on a Waters HPLC with a Supelco RP C18 column and a methanol:water (1:1) eluant. Absorbance was measured at 254 nm and 220 nm. The HPLC method was experimentally determined to have a TNT detection limit of 100 ppb. Dissolved organic carbon (DOC) and inorganic carbon (IC) analyses in liquid samples were performed on a Dohrmann DC-180 carbon analyzer and insoluble carbon was measured on a Dohrmann boat sampler unit. Carbon dioxide gas from the headspace was quantified on a Varian model 3700 gas chromatograph with a Supelco Carboxen-1000 column and TC detector. Nitrite, nitrate, and ammonium ion analyses were made using a Dionex ion chromatograph with conductivity detector and chemical suppression. Dissolved oxygen in the reactor was measured with an Orion model 860 oxygen meter and electrode. This probe has a detection limit of 0.1 mg/L O_2.

RESULTS AND DISCUSSION

In order to consider the efficacy of near UV light ($\lambda >340$ nm) to photocatalyze the destruction of TNT in TiO_2 slurries the first series of experiments was conducted to compare photocatalysis of TNT with direct photolysis. The reactor was initially aerated and the degradation was measured for 120 minutes. Figure 1 shows the percent loss of both TNT and dissolved organic carbon for these conditions. In each case the transformation of TNT was rapid and complete although the loss occurred at a higher rate under photocatalytic conditions. The initial pseudo-first order rate constant for the photocatalytic transformation of TNT was 4.2 hr^{-1} with essentially complete transformation (>99%) achieved within 60 minutes. With no catalyst present the initial pseudo-first order rate constant for TNT transformation was 1.2 hr^{-1} and approximately 90 minutes were required for complete transformation. The effect of the catalyst on the rate and extent of DOC (dissolved organic carbon) destruction was much more significant. More than 90% of the DOC was lost after 90 minutes with the TiO_2 present whereas less than 15% was lost within this same period in the absence of TiO_2. A carbon balance showed that the reduction in DOC which was measured in the photocatalysis of TNT was due to mineralization (oxidation to CO_2)

The results of the nitrite, nitrate, and ammonium ion analyses for these same experiments are presented in Figure 2. Concentrations are shown as a percentage of the total number of moles of nitrogen initially present as nitro groups on the TNT. With the catalyst present, the sum of the ammonium and nitrate ions was approximately 90% of total nitrogen after 120 minutes, further supporting the mineralization data shown in Figure 1. With no catalyst present the sum of nitrite, nitrate, and ammonium ions was approximately 8%. These results suggest that little ring cleavage occurred under direct photolytic conditions and that the observed 15% DOC loss was either due to the formation of a precipitate or mineralization of the methyl group on TNT. Oxidation of the methyl group has been reported for TNT photolysis by other researchers [10].

Under these photocatalytic conditions nitrite ion was rapidly oxidized to nitrate ion as has been observed in other work [11-13]. As shown in Figure 2, the nitrite ion concentration was

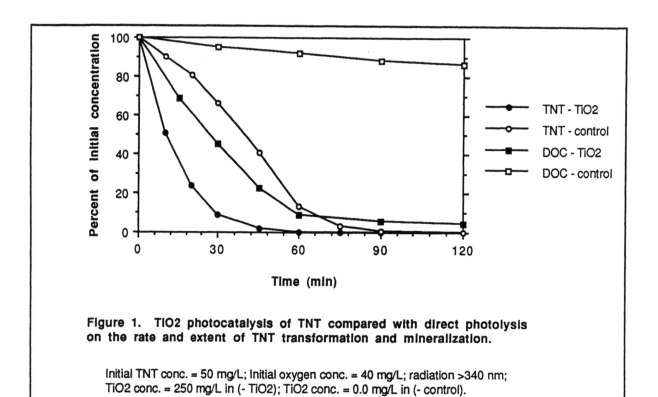

Figure 1. TiO2 photocatalysis of TNT compared with direct photolysis on the rate and extent of TNT transformation and mineralization.

Initial TNT conc. = 50 mg/L; Initial oxygen conc. = 40 mg/L; radiation >340 nm; TiO2 conc. = 250 mg/L in (- TiO2); TiO2 conc. = 0.0 mg/L in (- control).

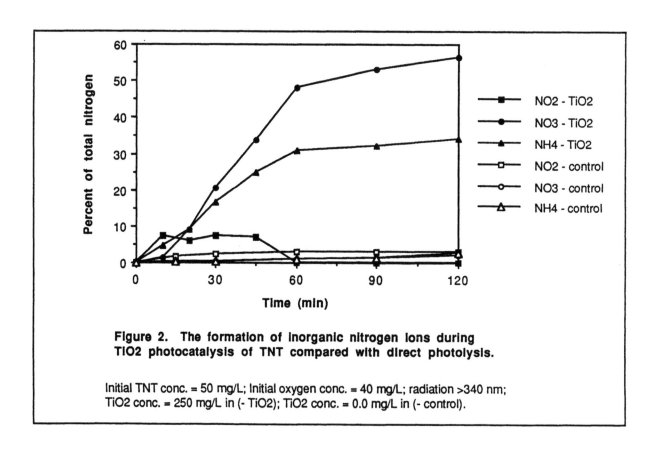

Figure 2. The formation of inorganic nitrogen ions during TiO2 photocatalysis of TNT compared with direct photolysis.

Initial TNT conc. = 50 mg/L; Initial oxygen conc. = 40 mg/L; radiation >340 nm; TiO2 conc. = 250 mg/L in (- TiO2); TiO2 conc. = 0.0 mg/L in (- control).

fairly constant for the first 45 minutes of the reaction as it was continuously formed and then oxidized to nitrate. As a result, the nitrate concentration increased steadily over this period. Between 45 and 60 minutes the rate of nitrite ion formation decreased as the reaction neared completion. Consequently, the nitrite ion concentration fell close to zero as the nitrate ion concentration approached its maximum.

The formation of ammonium ion during the photocatalytic degradation of a compound containing only nitro groups has been reported elsewhere [14]. However, in this work the final ammonium concentration for the photocatalytic degradation of TNT was approximately 35% of the total which gives a value of 0.67 for $[NH_4^+]/[NO_3^-]$. This is significantly higher than values reported for other compounds and may be due to differences in photocatalytic reaction conditions as well as to varying redox properties among the nitroaromatic compounds.

The possible conversion of nitrate ion to ammonium ion under these photocatalytic reaction conditions was assessed by repeating this experiment on a solution of 0.67 mM nitrate ion (from $NaNO_3$) and 1% tert-butyl alcohol as a hole scavenger at a pH of 5.0. Only 1.5% of the nitrate ion was reduced to ammonium ion after two hours. A similar experiment was performed with 0.67 mM $NaNO_2$ and 1% tert-butyl alcohol. In this case, the nitrite ion was completely oxidized to nitrate ion with negligible formation of ammonium ion. Additionally, oxidation of ammonium ion to nitrate ion was not detected when this photocatalytic experiment was run on a 0.67 mM solution of NH_4Cl at pH=5.0 for two hours. These results indicate that there was negligible conversion between ammonium ion and nitrate ion for the data shown in Figure 2.

These results and the fact that the production of inorganic nitrogen closely followed the course of DOC destruction indicate that nitrogen reduction in the photocatalytic system involved the nitro groups on the aromatic ring. This is consistent with observations and proposals of previous work. The production of aromatic nitroamines from the direct photolytic degradation of TNT has also been reported using both a germicidal lamp ($\lambda \approx 254$ nm) [15] and near UV ($\lambda > 280$ nm) [16,17] with identified nitroamines amounting to ≈ 10% by weight of the photolyzed TNT [17]. However, in a more recent study of TNT photolysis using a germicidal lamp seven aromatic byproducts accounting for approximately 80% of the DOC were identified and no aromatic amines were detected [10]. In these studies aromatic amines were detected only when ammonia was present in solution [10]. The mono-substituted aromatic nitroamines, 2-amino-4,6-dinitrotoluene and 4-amino-2,6-dinitrotoluene, were not detected in either the photocatalytic of photolytic experiments of this paper. It is thought that aromatic nitroamines were probably formed but were either not detectable by the HPLC method employed or did not accumulate to a sufficient degree to permit detection.

γ-RADIOLYSIS OF PHOTOLYZED TNT

The next phase of work was designed to indicate the extent to which direct photoreduction contributed to the formation of ammonium ion observed in the photocatalysis of TNT. Although only a small concentration of ammonium ion was detected during the non-catalyzed photolysis of TNT, it is not accurate to conclude that reduction of nitro groups to amine groups was negligible. The very limited degree of compound degradation which occurred under these conditions may have been insufficient to deaminate all reduced nitrogen. One route to deamination is proposed to proceed oxidatively via hydroxyl free radical attack at the site of the amine substitution resulting presumably in ring cleavage [14,18]. In a photocatalytic system hydroxyl radical formation is thought to be a primary event at the site of the positive hole so it is likely that deamination would occur readily in the catalyzed system.

γ-radiolysis of aqueous solutions sparged with N_2O is an effective method of generating hydroxyl radicals. In this work γ-radiolysis was utilized to expose photolyzed solutions of TNT to hydroxyl radical attack. Upon exposure to hydroxyl radicals, any aromatic amine groups that were formed by direct photoreduction of TNT should be deaminated and measurable as ammonium ion. Toward this end, a 50 mg/L solution of TNT was photolytically transformed by exposure to near UV radiation ($\lambda > 340$ nm) for a period of two hours. The concentration of TNT at this point was below the detection limit of 100 ppb. Afterwards, this photolyzed TNT solution and a control solution of TNT (50 mg/L) that had not been exposed to UV radiation were both sparged with N_2O and then exposed to γ-radiation generated by a ^{60}Co source. Comparing the amount of ammonium ion produced by the γ-radiolytic degradation of a photolyzed TNT solution with that of a non-photolyzed solution should give an indication of the degree of photoreduction which occurred during the photolysis of TNT.

Results of this work are shown in Figure 3 where the concentrations of nitrate and ammonium ion (again given as a percentage of total nitrogen initially present as nitro groups on the TNT) are shown for the γ-radiolysis of the photolyzed and non-photolyzed TNT solutions. After 480 minutes of γ-irradiation ammonium ion accounted for approximately 20% of the total nitrogen in the photolyzed TNT solution and 11% of the total nitrogen in the non-photolyzed TNT solution. The presence of ammonia in the non-photolyzed TNT solution indicates that not all the aqueous electrons were scavenged by N_2O yet the fact that the ammonia levels increase only over the first 60 minutes suggests that despite excess levels of N_2O, TNT reduction was competitive. The rest of the nitrogen in the non-photolyzed solution was recovered as nitrate ion with essentially 100% recovered. In the photolyzed solution the nitrogen recovered as ammonium plus nitrate was only 80% of total nitrogen so the possibility exists that other unidentified nitrogen species were present. Nitrite ion was rapidly oxidized to nitrate ion under these γ-radiolytic conditions and for this reason is not seen on Figure 3.

The potential for conversion between the different inorganic nitrogen ions during γ-radiolysis was assessed by γ-irradiation of salt solutions of nitrogen ions under the same conditions as the TNT solutions. No ammonium ion was formed by γ-irradiation of nitrite or nitrate ion but after a 480 minute γ-irradiation period approximately 20% of the initial ammonium ion was oxidized to nitrate ion. The results presented in Figure 3 have not been corrected to account for this oxidation of ammonium to nitrate because the degree to which this appears to have occurred does not affect the qualitative conclusions of this work.

These data clearly show that both oxidative and reductive reactions occurred during the γ-radiolysis of TNT. However, the higher ammonium ion concentrations generated during the γ-irradiation of photolyzed TNT indicate that the amount of photoreduction occurring during TNT photolysis is slightly greater than what is indicated by simply measuring ammonium ion concentration. Possibly in the range of 10% of total nitrogen is reduced by direct photolysis. This suggests that direct photoreduction contributes to the ammonium ion observed during the photocatalysis of TNT. Nevertheless, the significantly higher ammonium ion concentrations observed during photocatalysis of TNT (approximately 35% of total nitrogen) indicate that the predominant route of reduction in photocatalyzed systems is via interfacial reactions at the TiO_2 surface.

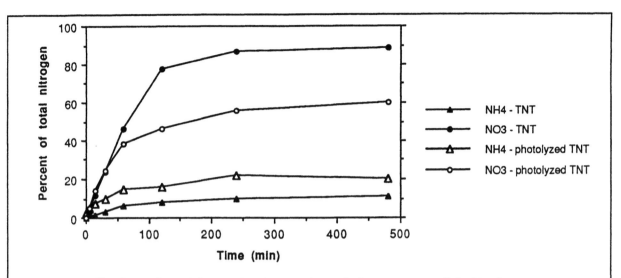

Figure 3. The formation of inorganic nitrogen ions during gamma radiolysis of a photolyzed TNT solution compared to gamma radiolysis of a TNT solution.

Initial TNT conc. in each solution = 50 mg/L; Photolyzed TNT solution was exposed to near UV radiation for two hours prior to radiolysis; Both solutions were sparged with nitrous oxide; Dose rate = 6.1 kgray/hr;

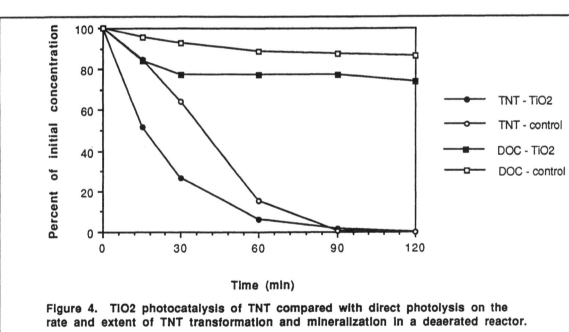

Figure 4. TiO2 photocatalysis of TNT compared with direct photolysis on the rate and extent of TNT transformation and mineralization in a deaerated reactor.

Initial TNT conc. = 50 mg/L; Initial oxygen conc. = 0.0 mg/L; radiation >340 nm;
TiO2 conc. = 250 mg/L in (- TiO2); TiO2 conc. = 0.0 mg/L in (- control).

PHOTOCATALYSIS OF TNT IN A DEAERATED REACTOR

The photocatalytic and photolytic experiments on TNT presented above were repeated under an atmosphere of nitrogen to explore further the capacity of TNT or its reaction by-products to serve as electron acceptors in place of oxygen. In the absence of either electron acceptors for conduction band electrons or electron donors for valence band positive holes, electron/hole recombination will occur in the photocatalyst preventing any photocatalytic reaction. The conventional picture of photocatalytic degradation of organic compounds is that molecular oxygen serves to sweep conduction band or trapped electrons while oxidative destruction occurs at the positive hole via either bound or free hydroxyl radicals. Under deaerated conditions, however, any enhanced degradation observed in the presence of the photocatalyst could occur only as a result of TNT or a reaction by-product serving to sweep electrons from the catalyst surface. These experiments allow consideration of the role of oxygen and that of direct electron transfer to nitroaromatic species at the semiconductor surface.

Figure 4 presents the percent loss of both TNT and dissolved organic carbon (DOC) vs. time for two experiments performed in the absence of oxygen and in the presence of near UV radiation ($\lambda > 340$ nm). As in previous experiments, one system contained and the other lacked TiO_2. In both experiments TNT was completely transformed but the rate was higher in the catalyzed experiment which had a pseudo-first order rate constant of 2.5 hr^{-1} as compared to a rate constant of 1.2 hr^{-1} for the experiment performed with no TiO_2. The removal of DOC was also greater in the presence of TiO_2. After 120 minutes approximately 26% of the DOC was lost in the photocatalytic experiment whereas only 13% was lost at this point in the experiment with no catalyst. These results suggest that TiO_2 did play a catalytic role in the absence of oxygen, thus, demonstrating that TNT or its reaction by-products can directly scavenge electrons from the catalyst surface.

It is of interest to note that in the absence of TiO_2 the rate and extent of TNT disappearance and mineralization under deaerated conditions was unchanged from that observed under oxygenated conditions. In the photocatalytic system, the elimination of oxygen had the effect of decreasing the initial pseudo-first order rate constant for TNT decay by 40% while the extent of mineralization was reduced by almost 70% at 2 hours. This suggests that oxygen is more effective at sweeping electrons than TNT and is needed for by-product destruction. Although it is expected that molecular oxygen is required for complete mineralization of organic carbon, the fact that mineralization was halted at such an early point (30 minutes) and at a stage where approximately 80% of the TNT had been transformed suggests that the transformation by-products of TNT were less effective electron scavengers. Photocatalytic reactions would then cease once suitable electron acceptors had disappeared.

Figure 5 is a graph of the evolution of nitrite, nitrate, and ammonium ions for these two experiments. The ion concentrations are again given as a percentage of the total number of moles of nitrogen initially present as nitro groups on the TNT and the sum of the inorganic nitrogen concentrations agrees approximately with the mineralization data shown in Figure 4. In the experiment with TiO_2 the predominant inorganic nitrogen ion was ammonium. This is indicative of the role of the nitro group as the primary electron acceptors. There was, however, a substantially lower ammonium ion concentration measured under deaerated conditions than was detected under oxygenated conditions. Moreover, the ammonium ion concentration reached a plateau at 60 minutes which coincided with less than 10% of the initial TNT concentration remaining. These observations are consistent with the proposal that the phototransformation products of TNT are

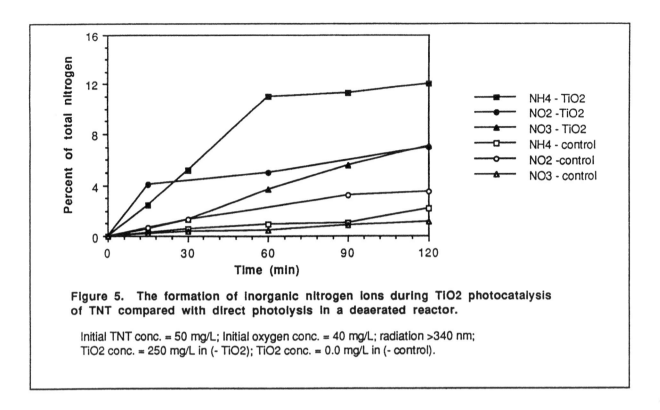

Figure 5. The formation of inorganic nitrogen ions during TiO2 photocatalysis of TNT compared with direct photolysis in a deaerated reactor.

Initial TNT conc. = 50 mg/L; Initial oxygen conc. = 40 mg/L; radiation >340 nm; TiO2 conc. = 250 mg/L in (- TiO2); TiO2 conc. = 0.0 mg/L in (- control).

less suitable electron scavengers than the TNT molecule itself. Such a result also suggests that photocatalysis was the primary source of hydroxyl radicals in this system and that once all the species available to sweep electrons from the catalyst surface had been transformed, there was no longer any effective mechanism by which hydroxyl radicals could be produced to deaminate the reduced aromatic species.

In the non-catalyzed experiment the distribution of inorganic nitrogen species was similar to that observed with the photolysis of TNT in an aerated reactor. Nitrite was again the predominant nitrogen ion. This result further indicates that the degradation pathways in this experiment were substantially different from those in the catalyzed one. Nitrite ion is also photochemically reactive and in the presence of radiation with $\lambda > 340$ nm excitation of nitrite ion leads to the formation of nitrogen monoxide and ultimately hydroxyl radical [19, 20]. Nitrite ion can also scavenge a hydroxyl radical to form nitrogen dioxide which disproportionates into nitrite and nitrate ions [23]. Such a mechanism could account for the formation of nitrate in these deaerated experiments. In the deaerated photocatalyzed experiment nitrite ion was much more stable than in the aerated one presented above. Previous work [11] found that the oxidation of nitrite ion (from $NaNO_2$) to nitrate ion in a deaerated TiO_2 photocatalytic reaction was negligible. However, in the experiment presented here TNT presumably serves as an electron acceptor allowing oxidation reactions at the positive hole to occur which may facilitate nitrite to nitrate oxidation.

CONCLUSIONS

These results have shown that TiO_2 photocatalysis using near UV radiation may be highly effective in the remediation of TNT contaminated waters. This process can achieve almost complete mineralization of TNT and it has the potential to be adapted to solar radiation which could offer an economic advantage over other advanced oxidation processes. The photocatalytic degradation of TNT appears to involve both oxidative and reductive steps. Ammonium ion accounts for approximately one third of the total nitrogen species produced under aerated conditions and appears to be formed predominantly via interfacial reactions at the catalyst surface and to a lesser extent by direct photoreduction. Data indicate that TNT is capable of sweeping electrons from the catalyst surface but not to the extent of fully reducing all aromatic nitro groups. Future work is expected to concentrate on illustrating degradation pathways through the identification of intermediate species and analyzing the specific role of the catalyst.

REFERENCES

1. Won, W.D., R.J. Hedkly, D.J. Glover, and J.C. Hoffsommer, (1974), Metabolic disposition of 2,4,6-trinitrotoluene, *Appl. Microbiol.*, vol. 27, pp. 513-516.

2. Keither, L.H., and W.A. Telliard, (1979), Priority pollutants. I. A perspective view, *Environ Sci Technol*, vol 13, pp. 416-423.

3. Tri-service environmental quality strategic plan program, developed by: The tri-service reliance joint engineers-environmental quality tech area panel, Draft program user review, F. Belvoir, VA, 7-11 December 1992.

4. Wujcik, W.J., W.L. Lowe, and P.J. Marks, (1992) Granular activated carbon pilot treatment studies for explosives removal from contaminated groundwater, *Environ Prog*, Vol. 11, No. 3, pp. 178-189.

5. Andrews, C.C., (1980), Photooxidative treatment of TNT contaminated wastewaters, Rep. No. WQEC/C-80-137 (AD-A084684), Weapons Quality Engineering Center, Naval Weapons Support Center, Crane, IN.

6. Matthews, R. W. (1990) Purification of Water with Near-U.V. Illuminated Suspensions of Titanium Dioxide, *Water Res,* Vol. 24, No. 5, pp. 653-660.

7. Mehos, M. S. and Turchi, C. S. (1993) Field Testing Solar Photocatalytic Detoxification on TCE-Contamnated Groundwater, *Env. Prog.,* Vol. 12, No. 3, pp. 194-199.

8. Schmelling, D., and Gray, K., "Photocatalytic Transformations and Mineralization of 2,4,6-trinitrotoluene (TNT) in TiO2 Slurries," submitted to *Environ Sci Technol.*

9. Murov, S.L., "Transmission characteristics of light filters," **Handbook of Photochemistry**, Marcel Dekker, Inc., NY, 1973, p. 97-103.

10. Hagen-LeFaivre, M. and Peyton, G. R. (1993) "Identification of By-products from Advanced Oxidative Remediation of Ground Water Containing Ordnance Compounds," presented at the 16th Midwestern Environmental Chemistry Workshop, University of Notre Dame, Notre Dame, IN.

11. Hori, Y., Nakatsu, A., and Suzuki, S., (1985), Heterogeneous photocatalytic oxidation of NO_2^- in aqueous suspension of various semiconductor powders, *Chem Lett*, pp. 1429-1432.

12. Hori, Y., Bandoh, A., and Nakatsu, A., (1990), Electrochemical investigation of photocatalytic oxidation of NO_2^- at TiO_2 (anatase) in the presence of O_2, *J. Electrochem. Soc.*, Vol. 137, No. 4, pp. 1155-1161.

13. D'Oliveira, J-C., Guillard, C., Maillard, C., and Pichat, P., (1993), Photocatalytic destruction of hazardous chlorine- or nitrogen-containing aromatics in water, *J. Environ. Sci. Health*, A28(4), pp. 941-962.

14. Low, G., McEvoy, S.R., and Matthews, R.W., (1991), Formation of nitrate and ammonium ions in titanium dioxide mediated photocatalytic degradation of organic compounds containing nitrogen atoms," *Environ Sci Technol*, Vol. 25, No. 3, p. 460-467.

15. Andrews, C.C., and Osmon, J.L., (1975), The effects of ultraviolet light on TNT in aqueous solutions, Rep. No. WQEC/C 75-197 (AD-B008175), Weapons Quality Engineering Center, Naval Weapons Support Center, Crane, IN.

16. Burlinson, N.E., Kaplan, L.A., and Adams, C.E., (1973), Photochemistry of TNT: Investigation of the 'Pink Water' problem, Rep. AD-769 670, Naval Ordnance Laboratory, White Oak, MD.

17. Kaplan, L.A., Burlinson, N.E., and Sitzmann, M.E., (1975), Photochemistry of TNT: Investigation of the 'pink water' problem. Part 2, Rep. No. NSWC/WOL/75-152 (AD-A020072), Naval Surface Weapons Center, White Oaks Laboratory, Silver Spring, MD.

18. Fox, M. A. (1983) "Organic Heterogeneous Photocatalysis: Chemical Conversions Sensitized by Irradiated Semiconductors" *Acc. Chem. Rex.*, Vol. 16, pp. 314-321.

19. Treinin, A., and Hayon, E., (1993), Absorption Spectra and Reaction Kinetics of NO_2, N_2O_3, and N_2O_4 in Aqueous Solution, *J Am Chem Soc*, Vol. 92, No. 20, pp. 5821-5828.

20. Alif, A., and Boule, P., (1991) Photochemistry and environment Part XIV. Phototransformation of nitrophenols induced by excitation of nitrite and nitrate ions, *J. Photochem. Photobiol. A: Chem*, 59, pp. 357-367.

TONY POWELL
BRIAN BUTTERS

Field Experience and System Enhancements of TiO$_2$ Photocatalytic Remediation Systems

ABSTRACT

TiO$_2$ Photocatalysis, an Advanced Oxidation Process (AOP), is quickly gaining recognition as an effective process for the remediation of organic contaminants in water and air. Its increased use, both in the bonded base and slurry form, have generated many data sets of contaminant reduction efficiencies. The data has been obtained by individuals possessing a widely varying knowledge base of TiO$_2$ Photocatalysis. They range from the novice level to the expert. Consequently, performance values of TiO$_2$ Photocatalysis for a predetermined set of conditions often have a <u>wide variance</u>. Variances in data sets occur when one or more key parameters have not been known or controlled during the tests. These key parameters directly affect the efficiency of the TiO$_2$ Photocatalytic process in a number of ways.

The intent of this paper is to look at some of these key parameters in detail for both air and water remediation and to outline a <u>standard of performance</u>. This paper attempts to relay a firm understanding of TiO$_2$ Photocatalysis based on extensive field experience gained over the last 20 months. The perspective of an operational point of view has been taken throughout the paper.

TEST STANDARDS - WATER REMEDIATION

In order to properly estimate operating costs for a TiO$_2$ Photocatalytic system, a firm understanding of the process and the parameters affecting TiO$_2$ Photocatalysis is required. This understanding is necessary in order to obtain both representable and repeatable data and thus allow one to analyze data sets (of both similar and different technologies) on equal terms. An operational understanding of TiO$_2$ Photocatalysis, and in particular of the key parameters which affect TiO$_2$ Photocatalysis, has been formed throughout the many field demonstrations and in-house treatability studies performed.

These parameters must be controlled or measured in order to effectively quantify TiO$_2$ Photocatalysis for the remediation of organic contaminants in water. For water remediation utilizing TiO$_2$ Photocatalysis, Reactor Performance (RP$_w$) is affected by the following key parameters.

Tony Powell P.Eng., Purifics Environmental Technologies Inc., 161 Mallard Road, Hyde Park, ON, N0M 1Z0
Brian Butters P. Eng., Purifics Environmental Technologies Inc., 161 Mallard Road, Hyde Park, ON, N0M 1Z0

$$RP_w = f(\text{Re, D.O., Initial TOC, Alkalinity, UV Flux, Electron Acceptors, } TiO_2)$$

Where:

Re - Reynolds Number depicts the flow regime (ie. laminar versus turbulent) inside the Photocatalytic cell. The Reynolds Number is more critical to a bonded catalyst based system since internal mixing provides the mode for mass transfer of organics to the bonded (immobilized) TiO_2 surface. Hence, the rate constant is tied directly to the Re. In a slurry process, the TiO_2 is suspended and thus provides greater mass transfer since the TiO_2 is in motion (provides translational and rotational movement) as well as the fluid.

D.O.- Dissolved Oxygen is critical for TOC removal and sustained conversion rates.

Background TOC Levels - Background organics will decrease the contaminant reduction rates. TiO_2 Photocatalysis is not selective to organic contaminants, thus all material will be oxidized rather than just the targeted compounds.

Alkalinity - A measurement of carbonate and bicarbonate ions. These ions are scavengers for the ·OH and will reduce the rate of contaminant reductions. Fluids with significant alkalinity levels (i.e. > 80 ppm) usually require pretreatment to remove these ions.

UV Flux - UV flux or light intensity is critical in order to supply the proper photon density to the TiO_2 particles. Various parameters affect the intensity and are listed further in the report. With slurry reactor systems, the percentage of TiO_2 can be adjusted to optimize the light density to TiO_2 particle ratio.

Electron Acceptors - Electron acceptors can increase the ·OH production by inhibiting the electron-hole recombination process. The effectiveness of electron acceptors are dependent upon the water quality, the fluid's chemical make-up, and the levels of contamination.

TiO_2 - The type of TiO_2 catalyst used has a direct effect on performance. TiO_2 particles are very stable and maintain its Photocatalytic activity effectively. For slurry applications, the amount of TiO_2 used in the process is the main operational parameter for maintaining performance. A peak level of performance is continually maintained by state of the art membrane separation technologies. Bonded TiO_2 systems lose performance over time due to loss of TiO_2 and thus require catalyst change-out on a scheduled basis.

PITFALLS

Other parameters are monitored but not necessarily a factor in the effectiveness of TiO_2 Photocatalysis as they may be dependent on the individual treatment system design. The following three parameters, which are called tracking indices for water remediation (TI_w), include pH, temperature, and voltage. These three indices are merely tools for controlling some of the seven key parameters which directly influence the effectiveness of the TiO_2 Photocatalytic process.

pH

Alkalinity is a good measure for the concentration of the carbonate and bicarbonate ions which, as discussed earlier are reaction "quenchers" as they compete for the 'OH. High concentrations of these ions can inhibit all contaminant reductions. One way of alleviating the problem is to lower the pH to approximately 4.0 where the ions are unstable and evolve as CO_2. At this point, contaminant reductions will proceed. Thus by controlling pH, the alkalinity is inherently controlled as well. Without knowledge of the effects of these ions on the Photocatalytic process, one may erroneously conclude:

- A very large pH dependence exists on TiO_2 Photocatalysis, or
- TiO_2 Photocatalysis was ineffective in treating that particular effluent (in which large concentrations of carbonate and bicarbonate ions exist).

TEMPERATURE

Temperature has no operational effect on the performance of TiO_2 systems. The slight increase in the reaction rate (k) with increasing temperature does not warrant any preheating of the fluid. However, inlet water temperatures can effect the UV Flux in some reactor designs and operationally may show half the performance of more refined designs.

VOLTAGE

As with temperature, low voltage to the power supply can also adversely affect light intensity to those systems employing traditional power supplies. This effect has been experienced first hand during an on-site demonstration in California. During the demonstration the voltage was measured to be so low that near "brown-out" conditions were experienced. However, modern power supplies are designed to compensate for voltage variances. Consequently, for a properly designed system, voltage fluctuations are no longer concerns for performance.

PERFORMANCE

When the seven parameters are controlled for optimal performance, a high standard of performance for TiO$_2$ Photocatalysis is maintained. Data of this quality can be compared to other data sets with confidence. One example of this kind of test is a 2,4-dichlorophenol destruction test which has typically been used to verify the performance of TiO$_2$ Photocatalytic Reactor Systems. A baseline performance standard is shown in Table I.

Table I: Performance Data

Contaminant	Inlet ppm	Outlet ppm	Treatment Time (s)	Removal Rate (ppm/min)	Power per m3 (kWh)	Power Cost ($/m3)
2,4-Dichlorophenol	10	3.5	45	8.7	2.3	0.14
2,4-Dichlorophenol	40	28.6	45	15.3	2.3	0.14
1,1,2-TCA	10	8.5	45	2.0	2.3	0.14

NOTE: The appropriate irreversible electron acceptor (chemical additive) can increase the removal rates by a factor up to 6 times. Power costs are based on a power cost of $0.06 per kWh.

COLOR/TURBIDITY

Color and turbidity are two parameters which strongly effect most light driven AOP technologies since a significant path length of light is usually required. However, the TiO$_2$ Photocatalytic technology is not affected by color and turbidity since the path length is millimeters instead of centimeters. This is a very large advantage when comparing this technology to other AOP technologies. TiO$_2$ Photocatalytic systems have demonstrated successful treatment on dyes (food and textile), and on highly turbid heavy water condensate from the nuclear industry. No effects were seen on TOC removal rates between turbid and non-turbid heavy water condensate.

ALL WATER IS NOT THE SAME

It is difficult to accurately estimate equipment size and operating costs for a specific application. It requires extensive field experience and a strong working knowledge of TiO$_2$ Photocatalytic Systems. Each water application differs in various ways which changes the Photocatalytic treatment efficiency. When the quality of the water varies, different chemical matrices are formed. This makes it very difficult to compare data between one application to another. Thus applications with "similar" targeted contaminant levels may not require the same level of treatment. For example, Table II lists two sites which have comparable inlet BTEX concentrations, but the level of treatment required to treat them are much different. The main reason for this is the different level of background organic contaminants.

Table II: Comparison of Different Quality Groundwater Samples

Parameter	Site A (Reference 2)		Site B (Reference 3)	
	Initial (ppb)	Final	Initial (ppb)	Final (ppb)
Benzene	127	ND	18.5	5.8
Toluene	110	ND	22.3	13.7
Ethylbenzene	180	ND	44.6	8.0
Total Xylenes	438	ND	425	122
MTBE	855	ND	ND	ND
Treatment Time	34 seconds		45 seconds	
Cost	$0.44/m^3		$0.48/m^3	

TOC Removal

Another water remediation application in which TiO_2 Photocatalysis proves quite effective is Total Organic Carbon (TOC) removal. TOC removal rates are dependent upon the organic contaminants to be mineralized. The work required for TOC removal is more intense than individual parameter destruction, thus TOC removal is usually performed in a batch mode to obtain the necessary treatment in an economical method. For TOC removal, dissolved oxygen becomes a more critical parameter due to the increased oxygen demand of mineralization. Consequently, oxygenation, aeration, or the addition of irreversible acceptors containing O_2 (i.e. H_2O_2) is required to maintain the optimal rates of TOC removal. The benefits of H_2O_2 addition are frequently over emphasized when it is compared to a base case TiO_2 system in which the dissolved oxygen has been depleted.

STATE-OF-THE ART

TiO$_2$ Photocatalytic Systems offer the following:
- Sustained unattended remediation with scheduled maintenance on an annual basis.
- Ability to perform remote diagnostics and control for system performance checks.
- Total turn-key systems which require limited manpower to install and operate.
- Operating pressures of up to 3.45 MPa (500 psi), with typical flow rates and pressure drops of 1 - 1000 GPM and 170 kPa (25 psi) respectively.
- Weather tight systems which can be installed either indoors or outdoors.
- No thermal controls or cooling requirements.
- Sustained operation without organic fouling or precipitation of Fe_2O_3.
- No operational impact on performance from color and turbidity.

189

AIR REMEDIATION

TEST STANDARDS

For air remediation utilizing TiO_2 Photocatalysis, Bonded TiO_2 systems have previously been used. For air remediation there are fewer key parameters as with water remediation which affect the TiO_2 Photocatalytic process. The reactor performance (RP_A) for air remediation utilizing TiO_2 Photocatalysis contain the following key parameters.

$$RP_A = f(Re, R.H., UV flux, TiO_2)$$

Where:

R.H. - Relative Humidity is an important factor for optimal performance. For typical applications such as air stripping, soil vapor extraction or plant air emissions, the typical RH levels in these applications do not inhibit the efficiency of the TiO_2 Photocatalytic process. This is important since an air drying pretreatment step is <u>not required</u> prior to remediation.

Re, UV Flux and TiO_2 - These parameters are important as with water remediation.

As well, the RP_A has been demonstrated not to be a function of temperature over a range of 0 - 80^{0}C (32 - 180^{0}F) for a sustained basis. The TiO_2 Photocatalytic process is an ambient temperature process, which is beneficial for applications of flammable gases.

STATE-OF-THE ART

A standard of performance for air remediation utilizing TiO_2 Photocatalysis is created when a firm understanding of the effects of the above key parameters are understood. Sustained field experience has demonstrated:

- Destruction of basic VOCs in only fractions of a second.
- TCE and PCE remediation from soil vapor extraction systems and air stripper columns at the 100 - 1000 ppm(v) range.
- Sustained TCE and PCE remediation with no phosgene production.
- Destruction of CCl_4.
- No operator requirement.
- In general, fixed TiO_2 Photocatalytic systems have been shown to handle TCE, PCE, and basic VOCs (levels around 500 ppm(v)) with removal efficiencies of 95 - 99% plus. The path lengths required to obtain this level of treatment are typically 1.6 m with pressure drops near 50 cm (20") water column.

OTHER FIELD EXPERIENCE GAINED

IN-SITU CLEANING FOR IRON REMOVAL FROM TiO$_2$

Dissolved iron (Fe^{2+}) in water, typically present in ground water, oxidizes very slowly in TiO$_2$ Photocatalytic systems creating Fe_2O_3. Without any pretreatment, the iron oxide will precipitate on the TiO$_2$, and thus will reduce the number of active sites for contaminant reductions. For bonded TiO$_2$ systems in which the inlet water contains dissolved iron, acid injection as a pretreatment is required to keep the iron from oxidizing. If this is not done, the system must be taken off-line periodically and continually recycled with an acid solution to remove the precipitated iron oxide. Both modes of iron treatment have been performed. The rate of the off-line acid recycle treatments is directly related to the inlet iron levels.

Slurry TiO$_2$ systems can pretreat the inlet water stream with acid as well, or they can employ acid washing of the TiO$_2$ concentrate from the membrane separation unit (ie. acid can be pumped into the TiO$_2$ concentrate to remove the iron oxide on a timed basis). This method would use less acid and can be performed unattended.

TECHNOLOGY COMPARISONS

BONDED, SLURRY AND OTHER TiO$_2$ PHOTOCATALYTIC PROCESSES

TiO$_2$ Photocatalytic Systems can be found in four basic operational modes as shown in the matrix below:

Slurry	Bonded	
X	X	Artificial Light
X	-	Solar Light

Ultraviolet light can be obtained naturally from the sun or generated artificially by electrically powered lamps. Systems using both light sources have been developed. Solar driven systems require large collector arrays on the order of several hundred to several thousand square feet. Solar systems are dependent upon the weather and can only operate at peak performance during the mid daylight hours. On the other hand, lamp driven system have a very small footprint and are available on demand at the flick of a switch.

Work has been performed by other parties to compare the effectiveness of the bonded TiO$_2$ Photocatalytic process to a slurry TiO$_2$ Photocatalytic process. Before they can perform such a comparison, they must understand exactly what they want to compare (i.e. the effectiveness of bonded TiO$_2$ Photocatalysis versus slurry TiO$_2$ Photocatalysis, or effectiveness of reactor designs). All to often unoptimized systems are used for such purposes. Thus, it turns out that a comparison between reactor designs is being compared rather than the process itself.

However, Measurement Technologies Inc. has compared various immobilized TiO$_2$ Photocatalytic reactor designs to the slurry process and has concluded that the slurry process outperformed all the immobilized designs by a factor of three[1].

The major advantage of slurry over bonded TiO$_2$ Photocatalytic Systems is the increased mass transfer obtained with the slurry design. In a fixed TiO_2 system the only mobile phase is the fluid. On the other hand, with the slurry system the TiO_2 is mobile as well, which typically doubles the level of mass transfer. As well, because the TiO_2 is mobile, rotational motion is available as well as translational motion; therefore, if an organic molecule adsorbs to a TiO_2 particle, the colloid can rotate to the direction of the light source. A bonded TiO_2 design will not permit this. With the bonded TiO_2 system, the organic contaminant <u>must</u> adsorb to the same side of the TiO_2 particle as the light source for oxidation to occur.

The slurry process must separate the TiO_2 from the fluid before discharging. This is accomplished by proven membrane technologies, which typically require only 20 - 25 psi of pressure head to operate. These are low maintenance units which operate unattended.

OTHER TECHNOLOGIES

The most effective measure for technology comparison are direct operating costs ($/ volume of fluid treated) and maintenance costs. Breaking out consumable costs (i.e. electrical, H_2O_2), operating and maintenance cost allocation, allows one to compare "apples to apples" when the technologies and processes are different. Capital costs can be misleading since a system can be amortized at varying rates and over different time periods. Similarly, capital cost varies with design, life, and operating conditions.

REFERENCES

1. O'Neil, T., P. Stricker, R. Burke, 1992. "Design and Fabrication of Prototype Solar Receiver/ Reactors." Report for National Renewable Energy Laboratory, Measurement Technology, Seattle.

2. Data Supplied by US Air Force Base - Tyndal, 1992.

3. Data Supplied by the State of Maine, Department of Environmental Protection, 1992.

ULICK STAFFORD
KIMBERLY A. GRAY
PRASHANT V. KAMAT

Photocatalytic Oxidation of 4-Chlorophenol on Titanium Dioxide: A Comparison with γ-Radiolysis

ABSTRACT

To gain useful insight into the mechanistic details of the TiO_2 photocatalyzed oxidation of halogenated organic compounds, the intermediates produced during the photocatalytic degradation of 4-chlorophenol (4-CP) have been compared with those produced during γ-radiolysis. Photocatalytic degradation of 4-CP produces aromatic intermediates consistent with γ-radiolytic hydroxyl radical oxidation, but the distribution of the intermediates differs. The surface area of TiO_2 has an important influence on the intermediate distribution suggesting that the presence of surface influences the reaction pathway. The course of photocatalytic transformation of 4-CP involves a combination of hydroxyl radical oxidation, direct electron transfer and surface chemical reactions contributing to the disappearance of 4-CP and its reaction intermediates in TiO_2 slurries.

INTRODUCTION

The photocatalytic destruction of many organic compounds in aqueous systems using titanium dioxide as a photocatalyst has been demonstrated [1]. This method, which may provide an alternative method for water and wastewater treatment, is facilitated when charge separation is induced in a large band-gap semiconductor material (e.g. TiO_2) by excitation with ultra-bandgap light. The electron-hole pair so formed, can then migrate to the catalyst particle surface and participate in interfacial redox reactions.

Halogenated aromatic compounds are a principal class of environmental pollutants. For this reason the photocatalytic degradation of the model compound, 4-chlorophenol (4-CP), has been studied extensively [2-11]. However, significant variations in aqueous phase intermediate concentrations have been reported during the photocatalytic degradation of 4-CP. In some photocatalytic studies hydroquinone (HQ) has been found to be the aromatic intermediate of greatest concentration [3,4,8], while in others 4-chlorocatechol (4-CC) was predominant [6,9,11]. The possible reasons for these differences have not been explained satisfactorily. Aromatic intermediate identity and concentration are of environmental concern, because the goal of treatment is to produce innocuous products and in some cases intermediates are more toxic than the primary substrate. Furthermore, knowledge of the destruction pathway and the kinetic details of chemical transformations are necessary in order to design photocatalytic systems that

Ulick Stafford, Department of Chemical Engineering, University of Notre Dame, Notre Dame IN 46556

Kimberly A. Gray, Department of Civil Engineering and Geological Sciences, University of Notre Dame, Notre Dame IN 46556

Prashant V. Kamat Notre Dame, Radiation Laboratory, University of Notre Dame, Notre Dame IN 46556

either achieve complete pollutant destruction or that might be integrated into a treatment train that combines various processes (e.g., biodegradation) to achieve complete destruction.

Radiolytic techniques are used to elucidate radical reaction mechanisms, and several radiolytic studies pertaining to the mechanistic details of TiO_2 photocatalysis have been reported [12-14]. In a recent study we used radiolytic techniques, both γ-radiolysis and pulse radiolysis to establish the steps in the degradation of 4-CP by free radical attack [11].

In this study the free radical pathways previously established for 4-CP degradation using γ-radiolysis [11] are compared with the photocatalytic mineralization of 4-CP in TiO_2 suspensions. The effects of increased TiO_2 loadings and light intensity on intermediate concentrations during photocatalysis experiments are examined, in order to determine their role in the degradation pathway.

EXPERIMENTAL

Full experimental details have been reported elsewhere [11]. 4-Chlorophenol (4-CP), 4-chlorocatechol (4-CC), 4-chlororesorcinol (4-CR), phenol, sodium azide and *tert*-butyl alcohol were reagent grade and used without further purification. Hydroquinone (HQ) was recrystalized from water/ethanol prior to use, and benzoquinone was purified by sublimation. Fumed titanium dioxide (P25, BET surface area - 50 m^2/g ± 5) was obtained from Degussa Corporation.

Photocatalysis was studied by illuminating reaction mixtures in an annular photoreactor (800 ml capacity) using either a 450 W medium pressure Hg lamp or an 8 W black lamp (λ_{max} 350 nm) within a borosilicate glass thimble, and is fully described elsewhere [15]. The reaction temperature was maintained at 28°C±0.5 when using the 450 W lamp and at 25°C±0.5 when using the 8 W lamp. Dissolved oxygen concentration, pH, and temperature were monitored constantly. Samples for HPLC analysis were filtered to remove the catalyst particles.

Samples were irradiated in HPLC sample vials in a Gammacell 220 $_{60}$Co γ-ray source (Atomic Energy of Canada, Ltd.). The source output was measured using Fricke dosimetry (1mM $FeSO_4.7H_2O$, 0.4M H_2SO_4) to be 8.68 Gy/min $\pm$ 0.40 [16].

HPLC using a reverse phase C-18 column was used to analyze for 4-CP and aromatic intermediates. A water-methanol eluant (60:40 with 1ml/l of acetic acid) were used. Absorbance at 280 nm was used to measure the concentrations of 4-CP and aromatic intermediates (except benzoquinone whose concentration was measured at 254 nm).

RESULTS

PHOTOCATALYSIS

Photocatalytic degradation of 250 μM 4-CP solutions was carried out with TiO_2 catalyst loadings of 0.025-1.00 g/l under UV-irradiation. The concentrations of 4-CP and oxygen and the reaction intermediates, HQ and 4-CC, for experiments conducted with the 450 W lamp are presented in Figure 1. A drop in the pH (from pH 6 to pH 4) was seen as a result of HCl formation. HQ and 4-CC are the major reaction intermediates in the photocatalytic oxidation process, but traces of benzoquinone, 4-chlororesorcinol, and 1,2,4-trihydroxybenzene were also detected. The rates of 4-CP degradation are similar at all except the lowest TiO_2 loadings when much of the light is transmitted through the slurry in the annular reactor (Table I), indicating that at a certain catalyst loading light becomes limiting. Beyond this limit [TiO_2] has little effect on 4-CP disappearance. In contrast, TiO_2 surface concentration exerts a significant effect on the maximum amount of 4-CC measured and on the rates of 4-CC production and destruction. The rate of oxygen consumption increased with increasing catalyst loading. Smaller differences are observed in the maximum HQ concentration as a function of catalyst loading. The [HQ] maxima

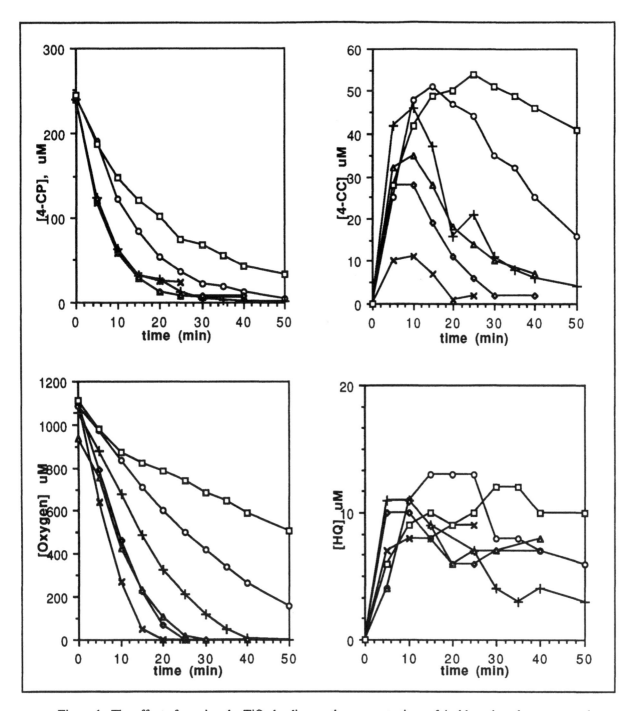

Figure 1. The effect of varying the TiO$_2$ loading on the concentrations of 4-chlorophenol, oxygen, and intermediates, during the photocatalytic degradation of 4-CP using a 450 W medium pressure Hg lamp. The TiO$_2$ loadings used were 0.025 g/l (□), 0.050 g/l(O), 0.125 g/l (+), 0.250 g/l(△), 0.500 g/l (◇), and 1.000 g/l (×).

are at approximately 10 μM which is much less than most 4-CC maxima. The one exception is at the highest catalyst loading (1.0 g/l) where the maximum concentrations of 4-CC and HQ are similar. In these experiments, as shown in other work [10,17], the rate of HQ disappearance is more sensitive to O$_2$ concentration than is that of 4-CC, and in the reaction at high catalyst loading in which all the oxygen is depleted rapidly, HQ concentration remains high. A summary of the maximum intermediate concentrations is presented in Table I, and similar data for a previously reported series of experiments conducted with the 8 W black lamp are presented in Table II [11].

γ-RADIOLYSIS

Results of the γ-radiolytic degradation of 4-CP are reported in detail elsewhere [11]. A summary is included for comparison with the photocatalytic results reported above. Yields in

TABLE I - TiO$_2$ PHOTOCATALYTIC DEGRADATION OF 4-CP USING A 450 W Hg LAMP [11]

TiO$_2$ concentration (g/l)	Transmittance through slurry T	Maximum 4-CC concentration [4-CC]$_{max}$ (μM)	Maximum HQ concentration [HQ]$_{max}$ (μM)	$\frac{[HQ]_{max}}{[4-CC]_{max}}$
0.025	0.71	54	12	0.22
0.050	0.57	51	13	0.25
0.125	0.23	46	11	0.24
0.250	0.08	35	11	0.29
0.500	0.04	28	10	0.36
1.000	0.00	11	9	0.82

TABLE II - TiO$_2$ PHOTOCATALYTIC DEGRADATION OF 4-CP USING AN 8 W BLACK-LAMP [11]

TiO$_2$ concentration (g/l)	Transmittance through slurry T	Maximum 4-CC concentration [4-CC]$_{max}$ (μM)	Maximum HQ concentration [HQ]$_{max}$ (μM)	$\frac{[HQ]_{max}}{[4-CC]_{max}}$
0.050	0.48	43	10	0.23
0.125	0.18	34	11	0.32
0.250	0.05	28	10	0.36
0.500	0.00	23	9	0.39
1.000	0.00	8	8	1.00

TABLE III - SUMMARY OF γ-IRRADIATION RESULTS [11]

Initial 4-CP concentration [4-CP]$_0$ (μM)	Major radical	pH	G(-4-CP)[i]	G(4-CC)[i]	G(HQ)[i]
244	•OH	3.0	3.3±0.3	1.9±0.2	0.3±0.2
247	•OH	6.1	3.6±0.3	2.1±0.2	0.3±0.2
247	•OH	9.1	2.7±0.3	-	-
65	•OH	6.0	3.3±0.3	1.5±0.2	0.5±0.2
983	•OH	6.6	3.6±0.3	1.8±0.2	0.6±0.2
1936	•OH	6.3	2.9±0.3	1.9±0.2	0.5±0.2
247	•N$_3$	6.0	1.9±0.3	-	-
1000[ii]	e$_{aq}^-$	6.0	0.8±0.3[ii]	-	-

i. G(-4-CP) is the yield of degradation of 4-CP. G(4-CC) and G(HQ) are yields of formation of 4-CC and HQ. G() is the number of molecules reacting per absorption of 100 eV of energy.

ii. Under these reducing conditions phenol was the major intermediate formed, G = 0.4±0.03.

196

terms of 4-CP disappearance and formation of intermediates are measured in terms of the G value, which is the number of molecules reacting per 100 eV of absorbed energy [18] and are presented in Table III. 4-CC and HQ, with traces of 4-chlororesorcinol, benzoquinone, 1,2,4-trihydroxybenzene, and other as yet unidentified trace compounds were formed when 4-CP was oxidized by hydroxyl free radicals in acid conditions, pH 3-6. Over this pH range the rates of 4-CP degradation and 4-CC and HQ formation vary little. In these mildly acidic conditions, ~60% of the 4-CP is converted initially to 4-CC and ~10% to HQ, during the first 10 minutes of irradiation. However this leaves a balance of ~30% that is degraded via other intermediates, that are not aromatic and are probably short-lived in the oxidizing conditions. These observations are consistent with the proposed •OH adduct (4-chlorodihydroxycyclohexadienyl radical) pathways presented in Figure 2 [6,11]. No aromatic intermediates were detected when 4-CP was degraded in basic conditions, and the yield in terms of 4-CP disappearance was slightly less. When 4-CP concentration was varied over the range, 64-2000 µM, at pH's ~3, ~6, and 9.1, there was little variation in yield indicating a zero order rate dependence on [4-CP]. When the 4-CP was oxidized directly by azo radicals in a buffered (pH 6) 0.05 M azide solution, no aromatic intermediates were detected, indicating that the reaction path as a result of direct electron transfer is different from hydroxyl radical attack. Phenol was the only major aromatic intermediate formed during reductive degradation of 4-CP.

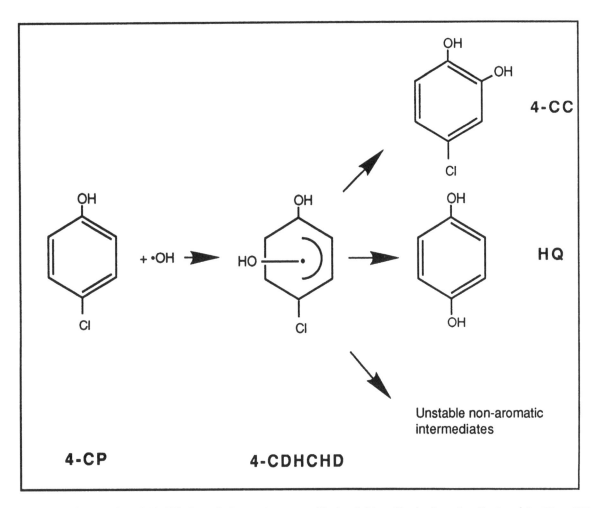

Figure 2. 4-Chlorophenol (4-CP) degradation pathways: oxidation initiated by hydroxyl radical attack. The •OH adduct, or 4-chlorodihydroxycyclohexadienyl radical (4-CDHCHD), subsequently decays via three different pathways forming either 4-chlorocatechol (4-CC), hydroquinone (HQ), or unstable aromatic intermediates [11].

DISCUSSION

The similarity between the observed aromatic intermediates in the hydroxyl free radical mediated γ-radiolysis experiments and the photocatalytic degradation experiments suggests that solution phase hydroxyl radical oxidation is a major reaction pathway in the photocatalytic degradation of 4-CP [11]. The absence of phenol as an intermediate in the photocatalytic studies suggests that solution phase reduction of 4-CP does not occur in TiO_2 slurries. However, there are significant differences in the relative concentrations of intermediates formed by the photocatalytic methods employed in the present study, in studies previously reported, and by γ-radiolysis suggesting that other important reactions occur during photocatalytic degradation of 4-CP [11].

CATALYST CONCENTRATION

The concentrations of 4-CP, intermediates, and oxygen vs. time in the photocatalytic degradation of 4-CP with variations of TiO_2 loading are shown in Figure 1. At higher catalyst loading ([TiO_2] ≥ 0.125 g/l), most of the incident radiation is absorbed so that little difference is observed in the initial rate of 4-CP degradation as a function of TiO_2 concentration (Table I). The concentration of 4-CC decreases considerably with increasing catalyst loading and may be due to a combination of factors. At higher loading less 4-CC may be produced because a greater proportion of 4-CP degradation may be taking place at the TiO_2 surface. Furthermore, 4-CC adsorbs strongly to the TiO_2 surface and as more surface is available a greater proportion of 4-CC would be adsorbed. HQ concentration was not seen to vary much as a function of TiO_2 loading. The low level of HQ measured relative to 4-CC at lower catalyst loadings results from a less preferred •OH substitution. As the amount of TiO_2 surface increases and more of the reaction takes place at the surface, HQ is the preferred surface reaction intermediate [10], and its production and desorption from the surface would compensate for decreased solution phase production. The initial rate of oxygen consumption increases significantly with increased TiO_2 loadings. These observations suggest that, despite the absence of change in the rate of 4-CP disappearance, it is being more rapidly *mineralized* at increased TiO_2 loadings, indicating more efficient use of the light.

LIGHT INTENSITY

Slightly different proportions of intermediate concentration were detected during experiments conducted using the 450 W medium pressure mercury lamp than using the 8 W black lamp for identical catalyst loadings. From Tables I and II the ratio of $[HQ]_{max}$ to $[4\text{-CC}]_{max}$ is consistently greater for the less intense lamp suggesting that a greater proportion of the degradation reactions occur on the TiO_2 surface under these conditions. The overall reaction rate, however, is faster with the more intense lamps and 4-CP disappearance occurred at least an order of magnitude more quickly. The greater proportion of reaction away from the surface in more intense light is probably due to a greater proportion of holes being scavenged by water to form hydroxyl radicals because the faster overall reaction rate decreases the concentration of other species adsorbed to the surface. Yet, preliminary figures suggest that the quantum efficiency of 4-CP disappearance is better for the weaker lamp ($\phi = 0.018$) than for the stronger lamp ($\phi = 0.008$) [15]. This difference in quantum yields is much less than predicted by the previously reported square root relationship [4].

OXYGEN CONCENTRATION

In Figure 1 it can be seen that an increase in TiO_2 concentration causes an increase in the rate of oxygen consumption. This suggests that an increase in TiO_2 concentration causes an increased rate of mineralization of the organic compounds. At high oxygen concentrations, O_2 consumption follows a zero order rate. At lower concentration the O_2 consumption rate becomes dependent upon oxygen concentration. In a previous study it was reported that photocatalytic reaction rates were dependent upon oxygen concentration up to saturation with a partial pressure of oxygen 60% of atmospheric pressure and independent of oxygen concentrations at higher levels [19]. The data in Figure 1 supports the observation that there is some point above which reaction rates are independent of oxygen concentration.

In the faster reactions, when the rate of oxygen consumption is faster that the rate of oxygen dissolution from the reactor head space, oxygen in solution becomes limiting. In such circumstances, e.g. 1.000 g/l in Figure 1, some 4-CP remains unreacted at the end, and HQ accumulates.

DISCREPANCIES IN INTERMEDIATE CONCENTRATIONS

Table IV includes values for intermediate concentrations relative to the initial 4-CP concentration for photocatalytic experiments reported here, in the literature, and from our γ-radiolysis results [11]. The concentrations and proportions of intermediates relative to the initial 4-CP concentration vary considerably. Much higher concentrations of 4-CC, were detected for the photocatalytic reaction at the lower catalyst loadings, and in all photocatalytic reactions [4-CC] was lower than during the γ-radiolysis of 4-CP solutions at pH 3-6 in conditions favoring •OH radical oxidation. In our photocatalytic experiments the concentrations of HQ were similar over the range of TiO_2 loadings. At the lower TiO_2 loadings the maximum concentration of 4-CC was four times greater than the maximum concentration of HQ, while at the highest TiO_2 catalyst loading (1 g/l) there was a 1:1 relationship between the maximum concentration of HQ and the maximum concentration of 4-CC.

Differences are also seen among the values of the previously reported studies (Table IV). The pH has been proposed as an explanation for these discrepancies [6]. γ-Radiolysis experiments in this study show that no aromatic intermediates were formed at pH 9. However, during the hydroxyl radical mediated degradation of 4-CP over the range pH 3-6 the nature and yields of intermediates are not dependent on the pH. All results listed in Table IV were conducted in unbuffered systems with initial pH 2-6, so pH does not adequately account for the discrepancies. We propose that reactor configuration and catalyst loading have a significant influence on the product distribution, and appear to more completely account for the discrepancies in the reported yields of reaction intermediates.

Significant concentrations of hydroquinone with only trace amounts of 4-chlorocatechol were detected in three previous studies [3,4,8]. The study by Al-Ekabi et al. was conducted in a tubular reactor illuminated from outside with TiO_2 immobilized on the inside wall [3]. It is possible that in this study all the light was absorbed by TiO_2 near the wall, and so few radical species were able to migrate from the solid-liquid interface. A dry powder study has shown that 4-CP adsorbed to the surface is degraded to adsorbed hydroquinone, the further degradation of which requires the presence of O_2 [10]. Therefore, the only reaction taking place in the study by Al-Ekabi et al. may be that of adsorbed 4-CP, reacting to form hydroquinone [3]. The study by Al-Sayyed et al. was conducted with a high catalyst loading in a deep reactor [4]. In this case it is likely that most light would be absorbed by TiO_2 particles near the wall. There may be significant adsorption of 4-CP in the dark region and subsequent reaction on the surface. Alternatively, much of the 4-CC formed may be adsorbed in this dark region, and thus was not detected. In the study by Yatmaz et al. the reaction was conducted in a falling film reactor, in

TABLE IV - MAXIMUM INTERMEDIATE CONCENTRATIONS FOR 4-CP PHOTOCATALYTIC DEGRADATION

Initial 4-CP concentration $[4\text{-}CP]_0$ (mM)	Maximum 4-CC fraction $\dfrac{[4\text{-}CC]m}{[4\text{-}CP]_0}$	Maximum HQ fraction $\dfrac{[HQ]m}{[4\text{-}CP]_0}$	Ratio of columns 2 and 3 $\dfrac{[HQ]m}{[4\text{-}CC]m}$	TiO$_2$ conc. (g/l)
0.250[i]	0.22	0.05	0.23	0.025
0.250[i]	0.04	0.04	1.0	1.000
0.250[ii]	0.19	0.04	0.21	0.050
0.250[ii]	0.03	0.03	1.0	1.000
0.100[iii]	0.183	0.063	0.34	0.5 g/l (P25)
0.155[iv]	trace	0.012	>5	2.0 g/l (P25)
0.063[v]	trace	0.039	>5	immobilized P25
0.250[vi]	-	0.200	>10	0.5 g/l
0.250[vii]	0.300[viii]	0.060[viii]	0.2	

i. Illuminated from the inside of annular photoreactor of pathlength 6 mm by a 450W medium pressure Hg lamp. Initial pH ~6.

ii. Illuminated from the inside of annular photoreactor of pathlength 6 mm by one 8W black lamp. Initial pH ~6 [11].

iii. Illuminated by six 8 W low pressure Hg lamps around reactor. Initial pH 2 [6].

iv. Illuminated from below by one 125 W medium pressure Hg lamp. Initial pH 3.4-6.0 [4].

v. Recirculated Tubular Pyrex reactor with TiO$_2$ immobilized on the inside illuminated from outside by six 15 W low pressure Hg lamps. Initial pH ~6 [3].

vi. Falling recirculated film illuminated by a 2.5 kW medium pressure Hg lamp. Initial pH ~6 [8].

vii. γ-Radiolysis [11].

viii. The last point measured was the maximum. It is possible that a higher maximum value would be recorded at later reaction times.

which the reaction mixture was briefly exposed to the light prior to a long period in the dark [8]. This time in the dark may allow equilibration of surface species with those in solution, not normally possible in constantly illuminated reactors. However, because of the high levels of HQ detected in this study, and the powerful lamp used with no apparent filtration of short-wavelength ultraviolet light ($\lambda < 290$ nm), it is possible that significant direct photodegradation of 4-CP also occurs [20].

IMPORTANCE OF SURFACE CHEMICAL REACTION

The concentrations of intermediates formed in semiconductor photocatalyzed degradation, especially at high catalyst loadings, are lower than those formed in γ-radiolysis. Also, the amount of HQ formed relative to the amount of 4-CC ($[HQ]_m/[4\text{-}CC]_m$) is greater in photocatalysis (Table IV). The concentration of HQ remains high once dissolved oxygen is exhausted. As the loading of TiO$_2$ particles in photocatalytic degradation reactions is increased the concentration of 4-CC decreases (Figure 1). Maximum 4-CC concentrations are also lower whereas HQ levels are unchanged in less intense illumination when a greater proportion of degradation is expected to occur on the surface. These results suggest that solution phase hydroxyl radical oxidation, although a major contributing factor, does not completely account for the observed product distribution. It also highlights the fact that surface reactions on TiO$_2$ particles play an important role in the photocatalytic degradation of 4-CP. Previous studies have shown that very small amounts of 4-CP are adsorbed by titanium dioxide from aqueous solution [6,7,21]. However, even such a small amount of adsorbed 4-CP may be important in long-term irradiation experiments, especially at high surface loadings. In an FTIR study of photocatalysis in a gas/solid system, we showed that 4-CP, chemisorbed to the TiO_2 surface, degraded under UV-irradiation to adsorbed HQ [10]. In aqueous systems the 4-CP is sparingly adsorbed, as it has to compete with water for adsorption sites. The ratio of [HQ] to [4-CC] (Table IV) is larger in conditions that favor more surface reaction and may be explained by a surface reaction in which hydroquinone is produced, by adsorption of 4-CC, or a combination of both. The long time in which catalyst particles are not illuminated would allow adsorption and desorption to occur [4,8]. In less intense light the lower concentration of 4-CC in solution suggests less solution phase •OH oxidation is occurring, and therefore a greater proportion reaction is occurring on the surface. The relatively high HQ concentration remaining after all oxygen was exhausted for experiments at high catalyst loadings may indicate that some HQ formed by such a surface reaction may subsequently desorb and accumulate when no oxygen is available to complete mineralization.

It has been shown that direct oxidation of several substrates at hole sites on the TiO_2 surface occurs [22]. In this case an electron is transferred directly from the substrate to a surface trapped hole. These conditions were simulated in solution when γ-radiolysis generated hydroxyl radicals were scavenged by azide to facilitate direct oxidation by azo radicals [11]. Under these conditions a reaction pathway without aromatic intermediates was predominant, which was initiated by electron transfer from 4-chlorophenoxide ions and 4-CP to the azo radicals. The lower concentrations of aromatic intermediates observed at increased surface in the photocatalytic degradation may be the result of an increased amount of degradation occurring via the direct electron transfer pathway.

ADSORPTION OF REACTION INTERMEDIATES

While adsorption of 4-CP to TiO_2 has been examined [6,7,21], little study has been made of the adsorption of 4-CC and HQ to the surface. Langmuir isotherm data show that 4-CC is adsorbed strongly to TiO_2 (K = 0.13 μM^{-1}, C_{sat} = 45 μ moles g^{-1}) [15]. An insignificant concentration of HQ is adsorbed in similar circumstances [15]. It is possible that low concentrations of 4-CC seen in some of the studies may be the result of unaccounted 4-CC that is adsorbed to the surface via a bidentate complex formation. Because more 4-CC than 4-CP is adsorbed to TiO_2, oxidation by valence band holes is likely to be more important for subsequent steps than it would be for the initial oxidation step of 4-CP.

INITIAL OXIDATION STEP

Most studies of TiO_2 assisted photocatalytic degradation have proposed that hydroxyl radical attack on the substrate is the first oxidation step. And in a study comparing TiO_2 photocatalysis with γ-radiolysis by Mao et al. it was shown that the oxidative degradation of chloroethanes proceeded by hydroxyl radical attack in TiO_2 photocatalysis [14]. However, in the same study it was shown that organic acids (trichloroacetic and oxalic) were oxidized primarily by valence band holes [14]. In a diffuse reflectance laser flash photolysis study of several compounds, including 2,4,5-trichlorophenol, over TiO_2 surfaces, no expected hydroxyl adducts were detected [22]. This suggests that in some surface promoted photocatalytic degradation reactions, oxidation occurs by direct electron transfer, and not by hydroxyl radical mediated attack. The pulse radiolysis of an aqueous solution of 4-CP and azide has shown that the 4-chlorophenoxide ion, and to a lesser extent, 4-CP, are directly oxidized by the azo radical to form 4-chlorophenoxyl radicals [11]. In γ-radiolysis under similar conditions, no aromatic intermediates are detected. The concentration of phenoxide ions is dependent upon the pH. At higher pH more of the 4-CP will be dissociated to the 4-chlorophenoxide anion increasing the likelihood of direct oxidation. The compounds mentioned above that are oxidized directly by valence band holes are stronger acids and as such are more dissociated than 4-CP at neutral pH (trichlorophenol, $pK_a \sim 7$, trichloroacetic acid, $pK_a = 0.51$, oxalic acid, $pK_a = 1.2$) [23].

It is evident from present studies of 4-CP degradation in TiO_2 slurries that oxidation occurs by a combination of hydroxyl radical and direct hole oxidation processes. For the poorly adsorbed 4-CP, the hydroxyl radical reaction will dominate when there is less surface available. The preferred site of •OH attack is the ortho position on 4-CP to yield mostly 4-CC ([HQ]/[4-CC] = 0.2 in γ-radiolysis). When the contribution from surface processes is promoted with increased TiO_2 loadings, more 4-CP degradation occurs on the surface to yield HQ. The adsorption of by-products such as 4-CC increases and the yield of detectable aromatic intermediates will decrease due to direct reaction with surface holes. Intermediate compound distributions are a function of the relative importance of these these phenomena: •OH attack, direct hole oxidation, and surface adsorption.

CONCLUSION

The complete photocatalytic degradation of 4-CP in TiO_2 slurries proceeds via a combination of solution phase hydroxyl radical oxidation and oxidation by valence band holes. The contribution of these reactions in general will depend on the surface loading, pH, and substrate properties such as pK_a and structure. Increased TiO_2 loadings favor surface reactions including direct valence hole oxidation and adsorption. The effect of increased catalyst concentration is to reduce the concentration of intermediates detected in solution, which may be desirable in an environmental application if the intermediates themselves are pollutants. It is also important to be able to predict product distributions in order to target or control the degradation reactions, allowing integration of a photocatalytic process with other treatment methods. Engineering parameters such as catalyst loading, light intensity, and reactor configuration are being further investigated to determine how much the rate of overall mineralization and quantum yield can be improved.

ACKNOWLEDGMENTS

The authors gratefully acknowledge the support of NSF [Grant No. BCS91-57948, K.A. Gray] and the Office of Basic Energy Sciences of the U. S. Department of Energy [P.V. Kamat] (this is contribution no. NDRL-3676 from the Notre Dame Radiation Laboratory). The authors thank the Center for Bioengineering and Pollution Control at the University of Notre Dame for the use of analytical equipment. The authors thank Degussa Corporation for the gift sample of TiO_2.

REFERENCES

1. (a) Schiavello, M., ed., 1988. Photocatalysis and Environment, Dordrecht:Kluver. (b) Serpone, N, and E. Pelizzetti, eds., 1989. Photocatalysis - Fundamentals and Applications, New York:Wiley. (c) Ollis, D.F., and H. Al-Ekabi, eds., 1993. Photocatalytic Purification and Treatment of Water and Air; Amsterdam:Elsevier.

2. Barbeni, M.,E. Praumero, E. Pelizzetti, E. Borgarello, M. Grätzel, and N. Serpone, 1984. "Photodegradation of 4-chlorophenol catalyzed by titanium dioxide particles". Nouv. J. Chim., 8:547-550.

3. (a) Al-Ekabi, H., and N. Serpone, 1988. "Kinetic studies in heterogenous catalysis. 1. Photocatalytic degradation of chlorinated phenols on aerated aqueous solutions over TiO_2

supported on a glass matrix". J. Phys. Chem., 92:5726-5731. (b) Al-Ekabi, H., N. Serpone, E. Pelizzetti, C. Minero, M.A. Fox, and R.B.Draper, 1989. "Kinetic studies in heterogenous photocatalysis. 2. TiO$_2$-mediated degradation of 4-chlorophenol alone and in a three component mixture of 4-chlorophenol, 2,4-dichlorophenol, and 2,4,5-trichlorophenol in air-equilibrated aqueous media" Langmuir, 5:250-255.

4. Al-Sayyed, G., J.-C. D'Oliveira, and P. Pichat, 1991. "Semiconductor-sensitized photodegradation of 4-chlorophenol in water". J. Photochem. Photobiol. A: Chem., 58:99-114.

5. (a) Matthews, R.W., 1986. "Photo-oxidation of organic material in aqueous suspensions of titanium dioxide". Wat. Res., 20:569-578. (b) Matthews, R.W., 1988. "Kinetices of photocatalytic oxidation of organic solutes over titanium dioxide". J. Catal., 111:264-272.

6. (a) Mills, A., S. Morris, and R. Davies, 1993. "Photomineralisation of 4-chlorophenol sensitised by titanium dioxide: a study of the intermediates". J. Photochem. Photobiol. A: Chem. 1993, 70:183-191. (b) Mills, A., and S. Morris, 1993. "Photomineralisation of 4-chlorophenol sensitized by titanium dioxide: a study of initial kinetics of carbon dioxide photogeneration". Ibid., 71:75-83.

7. Cunningham, J., and P. Sedlak, 1993. "Initial rates of TiO$_2$-photocatalyzed degradations of water pollutants: influences of adsorption, pH and photon-flux". in 1(c). pp. 67-81.

8. Yatmaz, H.C., C.R. Howarth, and C. Wallis, 1993. "Photocatalysis of organic effluents in a falling film reactor". in 1(c). pp. 795-800.

9. Sehili, T., P. Boule, C. Guyon, and J. LeMaire, 1989. "Photocatalysed transformation of chloroaromatic derivatives on zinc oxide. III: Chlorophenols". J. Photochem. Photobiol. A: Chem., 50:117-127.

10. Stafford, U., K.A. Gray, P.V. Kamat, and A. Varma, 1993. "An in situ diffuse reflectance FTIR investigation of photocatalytic degradation of 4-chlorophenol on a TiO$_2$ powder surface". Chem Phys. Lett., 205:55-61.

11. Stafford, U., K.A. Gray, and P.V. Kamat, 1994. "Radiolytic and TiO$_2$-assisted photocatalytic degradation of 4-chlorophenol. A comparative study". J. Phys. Chem, submitted.

12. Lawless, D., N. Serpone, and D. Meisel, D., 1991. "Role of •OH radicals and trapped holes in photocatalysis. A pulse radiolysis study". J. Phys. Chem., 95:5166-5170.

13. Terzian, R., N. Serpone, R.B. Draper, M.A. Fox, and E. Pelizzetti, 1991. "Pulse Radiolytic studies of the reaction of pentahalophenols with OH radicals: formation of pentahalophenoxyl, dihydroxypentahalocyclohexadienyl, and semiquinone radicals". Langmuir, 7:3081-3089.

14. Mao, Y., C. Schöneich, K.-D. Asmus, 1991. "Identification of Organic acids and other intermediates on oxidative degradation of chlorinated ethanes on TiO$_2$ surfaces en route to mineralization. A combined photocatalytic and radiation chemical study". J. Phys. Chem., 95:10080-10089.

15. Stafford, U., K.A. Gray, P.V. Kamat, and A. Varma, 1994. J. Catal., in preparation.

16. (a) Fricke, H., and E.J. Hart, 1966. in Radiation Dosimetry, Volume II, F.H. Attix, W.C. Roesch, eds. New York: Academic Press, Chapter 12. (b) McLaughlin, W.L., A.W. Boyd, K.H. Chadwick, J.C. McDonald, and A. Miller, 1989. Dosimetry for Radiation Processing, London:Taylor and Francis: London, pp. 144.

17. Vinodgopal, K., U. Stafford, K.A. Gray, and P.V. Kamat, 1994. "Electrochemically assisted photolysis. II. The role of oxygen and reaction intermediates in the degradation of 4-chlorophenol on immobilized TiO$_2$ particulate films". J. Phys. Chem, submitted.

18. Buxton, G.V., 1981. "Basic radiation chemistry of liquid water" in The study of Fast Processes and Transient Species by Electron Pulse Radiolysis, J.H. Baxendale, F. Busi, eds. Dortrecht:D. Reidel, pp. 241-266.

19. Augugliaro, V., L. Palmisano, A. Sclafani, C. Minero, and E. Pelizzetti, 1988. "Photocatalytic degradation of phenol in aqueous titanium dioxide dispersions". Toxicol. Environ. Chem., 16:89-109.

20. Boule, P., C. Guyon, and J. Lemaire, 1982. "Photochemistry and environment. IV-Photochemical behaviour of monochlorophenols in dilute aqueous solution". Chemosphere, 11:1179-1188.

21. Tunesi, S., and M. Anderson, 1991. "Influence of chemisorption on the photo-decomposition of salicylic acid and related compounds using suspended TiO$_2$ ceramic membranes". J. Phys. Chem., 95:3399-3405.

22. Draper, R.B., M.A. Fox, 1990. "Titanium dioxide photosensitized reactions studied by diffuse reflectance flash photolysis in aqueous suspensions of TiO$_2$ powder". Langmuir, 6:1396-1402.

23. Serjeant, E.P. and B. Dempsey, Ionisation Constants of Organic Acids in Aqueous Solution (IUPAC Chemical Data Series, No. 23), Oxford:Pergamon.

R. J. HILARIDES, K. A. GRAY
J. GUZZETTA
N. CORTELLUCCI
C. SOMMER

Degradation of Chlorinated Dioxins on Soil Using ^{60}Co Gamma Radiation: Considerations and Optimization

ABSTRACT

Soil contaminated with 2,3,7,8-Tetrachlorodibenzo-p-dioxin (TCDD) is a persistent environmental problem. Destruction of 2,3,7,8-TCDD on artificially contaminated soil using ^{60}Co gamma radiation has been demonstrated in our laboratory. A standard soil (EPA SSM-91) was artificially contaminated with 2,3,7,8-TCDD to 100 ppb. In the presence of water, non-ionic surfactant and high irradiation dose (800 KGy), greater than 90% TCDD destruction has been achieved. Many factors influence the effectiveness of destroying TCDD on soil including radiation dose, soil moisture content, and surfactant type, concentration, and mode of addition. The primary focus of the experiments presented in this paper was to evaluate the effects of soil amendments (water and surfactant) on the radiolytic destruction of TCDD. Byproducts and the reaction pathway are also discussed. It is shown that the reaction pathway for destruction of TCDD is a step-wise reductive dechlorination and that the initial reactions are relatively insensitive to free radical conditions (oxidative or reductive).

INTRODUCTION

BACKGROUND

Contamination by chlorinated dioxins is a persistent environmental problem to which there is no simple solution. While production from natural and artificial sources has resulted in widespread distribution of dioxin throughout the environment [1], there are approximately 500,000 tons of soil in the U.S. which are contaminated with dioxin to high levels requiring treatment [2]. Recent literature indicates that 2,3,7,8-tetrachlorodibenzo-p-dioxin (TCDD) remains a significant health threat suggesting that the highly restrictive regulatory standards

Roger J. Hilarides and Kimberly A. Gray, Department of Civil Engineering and Geological Sciences, University of Notre Dame, Notre Dame, Indiana 46556, USA

Joseph Guzzetta, Norma Cortellucci and Christopher Sommer, Occidental Chemical Corporation, Technology Center, 2801 Long Road, Grand Island, New York 14072, USA

currently in place are likely to remain in place [3-7]. Thermal treatments have been the most successful techniques for treating dioxin contaminated soil [2,8-9]. The most effective thermal technology is incineration; however, the capital and operating costs of incineration, coupled with poor public approval, limit its application [2,8]. Therefore, other effective and cost efficient treatment strategies must be identified and established.

Previous work has shown that ionizing radiation (e.g. UV-photolysis) can be used to degrade TCDD on soil in the presence of organic solvents or surfactant solutions [10-14]. Gamma radiation is another form of ionizing radiation and has been used to degrade TCDD in organic solvents [15]. However, there are no reports in the literature documenting degradation of TCDD on soils using gamma radiation. In our laboratory ^{60}Co gamma irradiation has been successfully employed to degrade chlorinated dioxins in a standard soil [16-17]. Furthermore, it has been shown through consideration of major operating and capital costs that a ^{60}Co gamma irradiation system is economically competitive with incineration and electron beam systems for the degradation of TCDD on soil [17]. The purpose of this paper is to consider more closely the influence of various factors on the extent and pathway of radiolytic destruction of TCDD.

RADIATION CHEMISTRY

Ionizing radiation can interact with matter by two main pathways to bring about chemical change. The first, and most commonly discussed, is **indirect radiation effects**. Indirect effect refers to the ionization of a solvent (e.g. radiolysis of water) which produces a suite of primary and secondary products (radicals) that diffuse through the solvent. These radicals then react with the compound of interest to bring about the desired chemical transformation or degradation. The products created in the greatest proportion from the radiolysis of water at neutral pH are the hydroxyl free radical ($^\bullet$OH) and aqueous electron (e^-_{aq}) [18]. Under these conditions the $^\bullet$OH and e^-_{aq} are created in a 1:1 ratio and are the primary reactive species in aqueous systems [19]. In soil/water systems the reactivity of the radicals created from radiolysis are difficult to predict due to the potential scavenging by components of the soil such as carbonates, the possible catalytic effects of the soil surface or trace metals, and the low solubility of contaminants such as TCDD.

The second pathway by which gamma radiation can interact with a compound is through **direct radiation effects**. Direct effects occur when the ionizing radiation deposits its energy in or very near the target compound. The target for the system described in these experiments is the TCDD molecule or possibly the TCDD molecule and the surfactant. The chemical change may occur through the ionization of the TCDD, either directly or more probably via a primary electron ejected from the deposition of energy in adjacent matter [20-21]. Statistical calculations using target theory [22] have been previously reported and show that the 37% dose (a statistical portion of TCDD molecules receiving a direct hit) may be as low as 800 KGy. These theoretical calculations coupled with the relative insensitivity of TCDD destruction to various reaction conditions observed in our laboratory are indication that the primary pathway of TCDD destruction in soil systems is through direct radiation effects [16].

206

OBJECTIVES

The results presented in this paper are part of a larger project that has the overall objective of developing an effective and efficient technique for destroying chlorinated dioxins on soil. TCDD destruction (transformation) in soils/slurries is a complex process and is influenced by many factors such as TCDD location in the soil matrix, soil characteristics, soil water content, co-contaminants, soil additives (e.g. surfactants), and soil-water-surfactant equilibration. Previous work identified the important parameters which affect TCDD destruction in soil [16]. The experiments presented herein are a continuation of previous work, and seek to further optimize the important parameters associated with water and surfactant addition such as concentration, sequence of addition, and equilibration time. Byproduct analysis is also presented from scavenger experiments. All results are considered and discussed as they relate to understanding the predominant pathway of destruction. Three key sets of experiments are discussed: 1) Water content optimization; 2) Surfactant addition optimization; and 3) Byproducts and the destruction pathway.

MATERIAL AND METHODS

SOIL CHARACTERIZATION

These studies were conducted using an artificially contaminated 'standard soil.' The 'standard soil' is from the U.S. E.P.A.'s Synthetic Soil Matrix Blending System (SSM-91). The SSM-91 soil constituents were established by an extensive review of soil characteristics of Superfund sites and eastern U.S. soils by the U.S. E.P.A. SSM-91 is a mixture of clay, silt, sand, top soil and gravel [23]. Further discussion of the soil characterization is provided elsewhere [16].

SOIL CONTAMINATION

Dioxin (2,3,7,8-TCDD) was obtained from Cambridge Isotope Laboratories. Artificial soil contamination to approximately 100 ppb TCDD was accomplished by a mixing procedure using hexane [16]. Homogenous contamination to approximately 100 ppb was verified and indicate that uniform soil contamination at a targeted level can be achieved successfully with this method.

TCDD ANALYSIS

Analyses for chlorinated dioxins were developed and conducted by the Dioxin Laboratory of Occidental Chemical Corporation, Technology Center. The dioxins were extracted from soil samples using a modified U.S. E.P.A. IFB series jar extraction [24]. The dioxins were isolated using an alumina column clean-up procedure. The extract was loaded onto a Hewlett Packard 5970B MSD coupled to a HP 5890 GC for HRGC/LRMS analysis, along with a 5 level calibration [25-26]. All results are surrogate recovery corrected. A more detailed discussion of this analytical procedure is provided elsewhere [16].

EXPERIMENTAL SETUP

The samples were prepared in 7.4 ml silanized borosilicate glass vials with teflon septa containing four grams of dry soil to which water and surfactant were added. Details of sample preparation are provided elsewhere [16]. Surfactant solutions have been shown to improve the destruction of TCDD in soil systems [13]. One non-ionic surfactant was selected for use in the experiments described herein based on successful use during screening experiments. The surfactant, Plurafac RA-40, is produced by BASF, Corp. and was obtained commercially. The optimum surfactant concentration is approximately 2% by weight of soil, as determined from previous experiments which considered both efficiency of TCDD destruction and cost [17].

GAMMA SOURCE

Naturally occurring and human-made radio-isotopes are excellent sources of gamma radiation. The experiments described in this paper used a 10,000 Curie Shepherd 109 ^{60}Co irradiator to produce gamma radiation. The gamma ray pair emitted from the spontaneous decay of ^{60}Co have energies of 1.17 and 1.33 MeV, and have high penetration lengths in materials of moderate density (soil/water slurries). ^{60}Co Gamma rays, therefore, will penetrate significantly further into a soil slurry than UV or visible light. The Shepherd 109 is a concentric, well type source that allows uniform dose distribution. Doses have been verified using National Bureau of Standards data and the results of Fricke dosimetry. The dose rates for this study ranged from 7.4 to 6.2 KGy/hr.

All samples were irradiated and analyzed in triplicate. TCDD destruction is a measure of the transformation of parent compound and was determined relative to a set of unirradiated controls that accompanied every group of samples replicating the conditions in the sample group.

RESULTS AND DISCUSSIONS

WATER CONTENT OPTIMIZATION

Soil moisture content has been identified as an important parameter in the destruction of TCDD adsorbed on soil. Previous work reported that 25% water content was the optimum value for TCDD destruction in soil [16]. These experiments, however, were conducted in the absence of surfactant and over a broad range of moisture content. Additional experiments have been conducted in order to determine if this optimum value is shifted in the presence of surfactant. Furthermore, by testing the extent of TCDD destruction at smaller increments of water addition a more careful look can be taken at indirect versus direct radiation effects. The dominant pathway of destruction is influenced by the relative proportions of solvent and solid [22]. Since indirect effects are proportional to the water

Figure 1 Water content optimization is shown for three irradiation doses (75, 150 and 450 KGy) with no surfactant present and at 150 KGy in the presence of 2% RA-40 surfactant. Data standard deviations are shown with error bars.

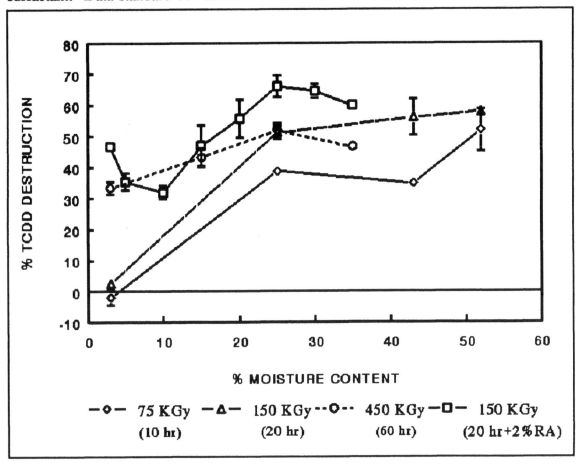

content of the irradiation volume, indirect radiation effects should become more significant with an increase in water content. However, due to the extremely low solubility of TCDD in water (0.2 ppb) [27-28], there may be a limit to the extent to which indirect effects increase as a function of water content.

Figure 1 is a plot of the percent destruction of TCDD as a function of soil water content with no surfactant at three doses (75, 150 and 450 KGy) and with 2% RA-40 surfactant at 150 KGy. For the three data sets with no surfactant, there is a general increase in the extent of TCDD destruction with increasing water content; however, beyond 25% there appears to be no significant gain in destruction. There are several competing effects taking place in this system. First, the relationship between direct and indirect effects is being altered by the addition of water. Increased water content should result in enhanced indirect effects by allowing production and diffusion of radicals; however, the mass being added (water) will not contain a significant amount of the TCDD thereby minimizing the enhancement. The addition of water may also enhance direct effects by providing an ionization medium adjacent to the TCDD for propagation of the direct effects. Second, the hydrophobicity of the TCDD may drive it further into the soil matrix making it less accessible to free radical attack. Still, the only increase in destruction is over the lower amounts of water addition (<25%), which suggests that indirect effects are not a major contributor at these low amounts of water, because most of the volume is still solid.

When surfactant is added, the heterogeneous system becomes even more multifarious. The role of the surfactant is very complex, though its enhancement of the radiolytic process is unequivocal [16-17]. In Figure 1, the enhancement in destruction is apparent when comparing the surfactant data to the no surfactant data at 150 KGy. In both cases, destruction improves with an increase in water content up to 25%. However, beyond 25% the increase is minimal, and in the presence of surfactant the extent of destruction appears to decrease. The presence of water in the soil-surfactant system enhances destruction by providing a medium for the surfactant to diffuse throughout the soil. After the initial enhancement up to 25%, addition of more water may then begin to dilute the surfactant, reducing its effectiveness. The apparent enhancement at very low water content (~3%) in the presence of 2% RA-40 is unexpected, yet it highlights the complex role that the surfactant plays in this process. The most important result derived from Figure 1, though, is confirmation that 25% water content is the optimum level for the destruction of TCDD on SSM-91 soil in the presence of 2% RA-40 surfactant.

SURFACTANT OPTIMIZATION

Despite the fact that the complex role of the surfactant is not yet fully understood, optimization of its use may be possible. Previous work showed that rates of TCDD destruction diminished at long irradiation times resulting in a residual TCDD level of 7 ppb. It was hypothesized that the surfactant was being degraded under the radiation flux. Therefore, in order to optimize the process and achieve destruction to a level less than 1 ppb TCDD, two experiments were conducted to see if improvement would occur through the method of surfactant and water application. The first experiment looked at water-surfactant addition sequence and equilibration times. The second investigated intermittent surfactant addition.

SEQUENCING

The data presented in Figure 2 show the results of experiments which tested the sequence of water and surfactant addition, and the effect of sample equilibration times following amendment addition and before irradiation. The bar graph shows the amount of TCDD remaining following irradiation, and the unirradiated controls are provided for reference. Sequences **A** and **B** are water addition with two equilibration times and no surfactant added. There is negligible difference between a one week equilibration (**A**) and a 24 hour equilibration (**B**). It had been previously hypothesized that the water addition would drive the TCDD further into the soil matrix, making it less accessible for radiolytic destruction. Over the time scale shown for this data, the negative impact of water equilibration is not observed.

It was previously established that surfactants improved the level of TCDD destruction and a minimum 24 hour equilibration period was required for the surfactant to be fully effective. Sequences **C**, **D**, and **E** all had 2% RA-40 surfactant added and the extent of

Figure 2 Water-surfactant sequencing experiment. Water addition to 25% (w/w) and surfactant addition to 2% using RA-40. Irradiation dose is 150 KGy. Standard Deviations are shown as error bars.

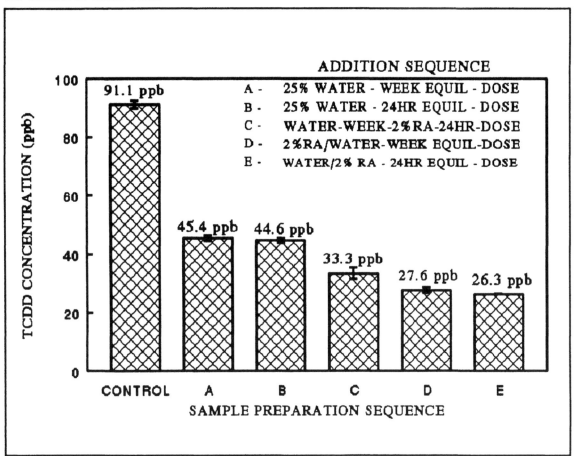

TCDD destruction while significantly improved over the no surfactant data is similar among them. Sequence **C** had the water added first, then a one week equilibration period, followed by surfactant addition, a 24 hour equilibration and then irradiation. This sequence of addition was slightly less efficient than adding the surfactant with the water, implying that there may be a small impedance to destruction efficiency by allowing the water and soil to equilibrate prior to surfactant addition. Sequences **D** and **E** had the surfactant and water added initially with no equilibration between their additions, followed by two different equilibration periods. These results show that the extent of TCDD destruction is not enhanced by increasing the equilibration period from 24 hours to one week. Finally, though there is negligible impact from water-soil equilibration in the no surfactant systems (**A** and **B**), there is some difference in destruction in surfactant systems due to water-soil equilibration (**C** and **D**). This further illustrates the important role that water has on surfactant efficiency for the radiolytic destruction of TCDD.

Figure 3 Intermittent surfactant addition (soil with 25% water content). (Note 1): The data with open circles are the intermittent addition up to 2% total surfactant, open boxes are with 2% added initially.

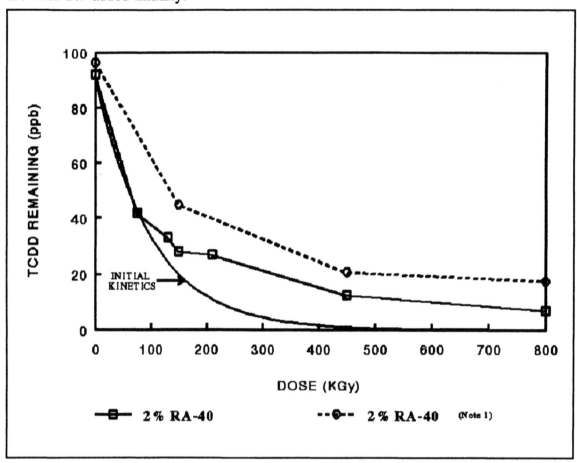

INTERMITTENT SURFACTANT ADDITION

Previous experiments identified the initial kinetics (1 to 110 KGy) of TCDD destruction on soil with 25% water content and 2% RA-40 surfactant, and these initial kinetics are shown with a single line in Figure 3. Based on the initial kinetics and the target theory calculations discussed earlier, it was expected that destruction to less than 1 ppb TCDD could be achieved with a dose of 800 KGy, but destruction begins to fall off around 100 KGy possibly due to surfactant destruction. It is hypothesized that intermittent surfactant addition might minimize the effect of surfactant destruction over the course of a long irradiation.

Intermittent surfactant addition was accomplished by adding 0.5% surfactant to sample vials initially and then three more times after each quarter of the total dose had been delivered to the sample, resulting in a total of 2% surfactant being added. This experiment was conducted at three doses (150, 450, and 800 KGy) and the results are compared to data where all the surfactant (2%) was added prior to irradiation. These results are shown in Figure 3 with the concentration of TCDD remaining being plotted as a function of radiation dose. Unfortunately this method of incremental surfactant addition failed to enhance the rate or extent of radiolytic destruction of TCDD. Surfactant destruction (modification) may still be the cause of the limited destruction but these results suggest that there is a threshold level of surfactant required between 0.5 and 2.0%. Since there are many possible intermittent addition scenarios, more experiments are required to determine an optimum practice. Direct analysis of surfactant destruction is also being investigated.

BYPRODUCTS AND RADIATION EFFECTS

Analysis of the irradiated soil samples for byproducts has closed nearly 90% of the mass balance at low irradiation doses (<150 KGy) and has shown the pathway of destruction of TCDD on soils to be through step-wise reductive dechlorination under normal reaction conditions [16]. Although these results and others suggest that primarily direct effects promote TCDD destruction, byproduct analysis of scavenger experiments was performed to provide further evidence that direct effects prevail over indirect effects in these soil systems.

Figure 4 Byproducts and reaction conditions (25% water and 150 KGy): **A** no surfactant, **B** 2% RA-40/N$_2$O (oxidative), **C** 2% RA-40/4% IPOH/N$_2$ (reductive), **D** 2% RA-40/N$_2$.

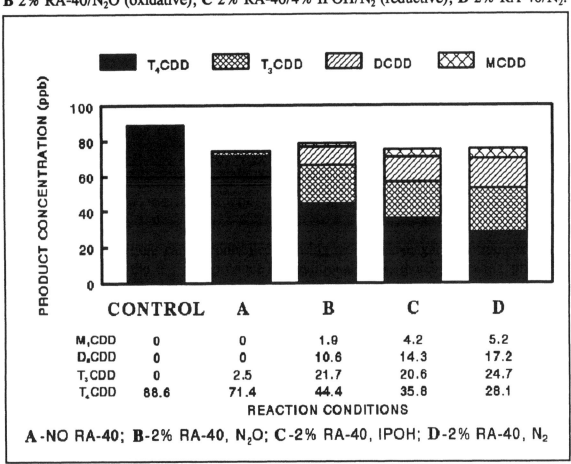

	CONTROL	A	B	C	D
M$_1$CDD	0	0	1.9	4.2	5.2
D$_2$CDD	0	0	10.6	14.3	17.2
T$_3$CDD	0	2.5	21.7	20.6	24.7
T$_4$CDD	88.6	71.4	44.4	35.8	28.1

REACTION CONDITIONS

A-NO RA-40; **B**-2% RA-40, N$_2$O; **C**-2% RA-40, IPOH; **D**-2% RA-40, N$_2$

Reaction conditions can be controlled by using radical scavengers. Nitrous oxide gas (N$_2$O) will scavenge the e$^-_{aq}$ and produce an additional $\cdot$OH, thereby creating primarily oxidative conditions. Alternately, alkaline isopropanol in conjunction with a N_2 gas sparge will scavenge the $^\bullet OH$ to produce a relatively unreactive radical species promoting e^-_{aq} predominance and reductive conditions. Samples which were sparged with N_2 gas will have conditions which are simultaneously reductive and oxidative because N_2 eliminates scavengers and the $^\bullet OH{:}e^-_{aq}$ is 1:1. Although reaction conditions in soils may not be controlled as easily as in solution experiments because of the complexity of the soil matrix, the use of these scavengers should have a significant impact on the TCDD destruction if indirect radiation effects are dominant.

The data presented in Figure 4 show the total poly-chlorinated dibenzo-p-dioxin (PCDD) concentration as total bar height, for varied reaction conditions after irradiation to 150 KGy. Each bar is comprised of the contribution from each of the lesser chlorinated dioxin byproducts and the TCDD remaining. For the data with no surfactant (A), lesser TCDD destruction (20%) occurred in comparison to conditions with surfactant (B-D) (50-70%) and only a small amount of T_3CDD (tri-chloro) was produced. Under primarily oxidative conditions (B) slightly less TCDD loss was observed in comparison to conditions C and D, but approximately 80% of the identified byproducts are attributed to step-wise reductive dechlorination. This demonstrates that the primary pathway for destruction is not through an indirect oxidative pathway. Under primarily reductive conditions (C), the destruction is not statistically different from that observed under oxidative (NO_2) conditions, although it is slightly greater. Finally, under simultaneously oxidative and reductive conditions (D) the greatest extent of TCDD transformation occurred with a byproduct distribution of lesser chlorinated dioxins, but once again it is not significantly different from the destruction of (B and C). These experiments nearly close mass balance and show definitively that the reaction conditions will not significantly affect the reaction pathway, thereby implying that the pathway is determined primarily by direct radiation effects.

CONCLUSIONS

This paper provides information concerning the effective use of ^{60}Co gamma radiation for the destruction of TCDD in soil. Appropriate application of water and surfactant will have a major impact on the rate and extent of TCDD destruction. Four major conclusions are derived:

1) The optimum water content was verified with surfactant present in the soil system. The maximum extent of destruction occurred near 25% water content, and confirmed previous results showing this to be an optimum level when no surfactant was present.

2) The sequence of water-surfactant addition as well as the equilibration times prior to irradiation can have an impact on TCDD destruction. The addition of surfactant enhances destruction in comparison to water only systems and no differences were found due to the order of water-surfactant addition or between a 24 hour or 1 week equilibration. A 24 hour equilibration in the soil-water-surfactant system is required though.

3) Intermittent surfactant addition may enhance the rate and extent of TCDD destruction by replenishing surfactant that was degraded under the radiation flux; however, intermittent

addition of 0.5% surfactant (up to a total of 2%) during irradiation was less effective than adding 2% surfactant initially suggesting that a threshold level is required. Future experiments will investigate other intermittent surfactant addition concentrations to identify an effective treatment scheme.

4) Byproduct and scavenger experiments effectively closed the mass balance and elucidated the reaction pathway for destruction of TCDD. The byproducts are primarily lesser chlorinated dioxins, therefore the reaction pathway is through reductive dechlorination. The relative insensitivity of the total destruction and byproduct distribution to the reaction conditions (oxidative or reductive), coupled with previous theoretical calculations using target theory, are strong indicators that TCDD destruction is occurring primarily through direct radiation effects.

The results presented are major steps in developing an effective and efficient process for the destruction of chlorinated dioxins adsorbed on soil. The predominant pathway of TCDD transformation appears to be via direct radiation effects and research continues to determine the optimum combination of parameters to achieve destruction below the 1 ppb TCDD level at doses in the 800 KGy range.

ACKNOWLEDGEMENTS

The authors thank Occidental Chemical Corporation for sponsoring this project, Occidental's Technology Center for their technical assistance, and Occidental's Dioxin Laboratory for their analytical support. We would also like to thank Notre Dame's (U.S. Department of Energy) Radiation Laboratory for the use of their ^{60}Co source, and especially Dr. Larry Patterson for many helpful and insightful suggestions.

REFERENCES

1. Berry, M.B.; C.E. Luthe and R.H. Voss, 1993. "Ubiquitous Nature of Dioxins: A Comparison of the Dioxins Content of Common Everyday Materials with That of Pulps and Papers." Environmental Science & Technology, 27:1164-1168.

2. U.S. Congress, Office of Technology Assessment, 1991. Dioxin Treatment Technologies-Background Paper, OTA-BP-O-93 (Washington, DC: U.S. Government Printing Office, November 1991).

3. Krukowski, John, 1992. "Special Report: Cracking Dioxin's Secrets." Pollution Engineering. January 15, 1992, pp. 19-21.

4. Hanson, David J., 1991. "Dioxin Toxicity: New Studies Prompt Debate, Regulatory Action." C&EN, August 12, 1991, pp. 7-14.

5. Health & Environment Digest, 1992. "Dioxins & Their Cousins: The Dioxin '91 Conference," 5(12):1-10.

6. Preuss, P.W. and W.H. Farland, 1993. "A Flagship Risk Assessment." EPA Journal, 1:24-26.

7. Stone, R.,1993. "Dioxin: Still Deadly." Science, 260:31.

8. Arienti, M., 1988. Dioxin Containing Wastes: Treatment Technologies. Park Ridge, NJ, Noyes Data Corp.

9. Edwards, N., 1992. "Assessing Dioxin Treatment and Disposal." Water Environment & Technology, 4:58-60.

10. Bertoni, G.; D. Brocco; V. Di Palo; A. Liberti; M. Possanzini; and F. Bruner, 1978. "Gas Chromatographic Determination of 2,3,7,8-Tetrachlorodibenzodioxin in the Experimental Decontamination of Seveso Soil by Ultraviolet Radiation." Analytical Chemistry, 50:732-735.

11. Crosby, D.G.; A.S. Wong; J.R. Plimmer; and E.A. Woolson, 1971. "Photodecomposition of Chlorinated Dibenzo-p-Dioxins." Science, 173:748-749.

12. Dougherty, Erika J.; A.L. McPeters; M.R. Overcash; and R.G. Carbonell, 1993. "Theoretical Analysis of a Method for in Situ Decontamination of Soil containing 2,3,7,8-Tetrachlorodibenzo-p-dioxin." Environmental Science & Technology, 27:505-515.

13. Exner, Jurgen H.; E.S. Alperin; A. Groen, Jr.; and C.E. Morren, 1984. "In-Place Detoxification of Dioxin-Contaminated Soil." Hazardous Waste. 1:217-223, 1984.

14. Kieatiwong, Somchai; L.V. Nguyen; V.R. Hebert; M. Hackett and G.C. Miller, 1990. "Photolysis of Chlorinated Dioxins in Organic Solvents and on Soils." Environmental Science & Technology, 24:1575-1580.

15. Fanelli, R.; C. Chiabrando; M. Salmona; S. Garattini; and P.G. Caldera, 1978. "Degradation of 2,3,7,8-Tetrachlorodibenzo-p-dioxin in organic solvents by gamma ray irradiation." Specialia, Experientia, 34:1126-1127.

16. Hilarides, Roger J.; K. Gray, J. Guzzetta, N. Cortellucci, and C. Sommer, 1993. "Radiolytic Degradation of 2,3,7,8-TCDD in Artificially Contaminated Soils." Submitted to Environmental Science & Technology, July 1993.

17. Hilarides, Roger J.; K. Gray, J. Guzzetta, N. Cortellucci, and C. Sommer, 1993. "Radiolytic Degradation of Dioxin on Soil: Optimal Conditions and Economic Considerations." In Press, Environmental Progress. November 1993

18. Buxton, G.V., 1987. "Radiation Chemistry of the Liquid State: Water and Homogeneous Aqueous Solutions." In Radiation Chemistry: Principles and Applications, Edited by Farhataziz and M.A.J. Rodgers, New York, VCH Publishers.

19. Klassen, N.V., 1987. "Primary Products in Radiation Chemistry." In Radiation Chemistry: Principles and Applications, Edited by Farhataziz and M.A.J. Rodgers, New York, VCH Publishers.

20. LaMarsh, J.R., 1983. Introduction to Nuclear Engineering, 2nd ed., Reading, MA., Addison-Wesley Publishing Co.

21. Swallow, A.J., 1973. Radiation Chemistry. New York, John Wiley & Sons. pp. 136-156.

22. Lea, D.E., 1955. Actions of Radiations on Living Cells. New York, Cambridge University Press.

23. Tabak, M.E.; W. Glynn; and R.P. Traver, 1991. "Evaluation of EPA Soil Washing Technology for Remediation at UST Sites." U.S. EPA Report.

24. IFB WA 84-A002, 1983. Dioxin Analysis, Soil Sediment Matrix Multi-Concentration, Selected Ion Monitoring (SIM) GC/MS Analysis With Jar Extraction Procedure. 9/15/83, Amendment I, 12/29/83.

25. EPA Method 1613: Tetra- through Octa- Chlorinated Dioxins and Furans by Isotope Dilution. July 1989.

26. EPA Method 8280, revision 0, dated September 1986

27. Chiou, Cary T. and M. Manes, 1990. "Comment on 'Temperature Dependence of the Aqueous Solubilities of Highly Chlorinated Dibenzo-p-dioxins.'" <u>Environmental Science & Technology</u>, 24:1755-1756.

28. Shiu, Wang Ying; W. Doucette; F. Gobas; A. Andren; and D. Mackay, 1988. "Physical-Chemical Properties of Chlorinated Dibenzo-p-dioxins." <u>Environmental Science & Technology</u>, 22:651-657.

LIXIONG LI
PEISHI CHEN
EARNEST F. GLOYNA

Pilot-Plant Validation of Kinetic Models for Supercritical Water Oxidation

ABSTRACT

The supercritical water oxidation (SCWO) process is an effective and environmentally attractive option for treating organic wastewaters and sludges. Currently, efforts are being made by several private companies and government agencies to commercialize the SCWO process for the treatment of domestic and industrial wastewaters and sludges, and for the destruction of the noxious organic components of nuclear and military wastes. However, there is a lack of design data based on pilot-scale kinetic studies.

While SCWO kinetic models, based on laboratory-scale data, were available for some simple compounds, the purpose of this pilot-scale study was to validate these models for selected compounds. Specfically, acetic acid, phenol, and n-octanol were studied using a 6.1-m long, concentric-tube reactor with 150-liter/hour (40 GPH) throughput. Process variables included temperature (385°C-440°C), feed concentration (acetic acid: 4 g/L-21 g/L; phenol: 0.59 g/L-0.77 g/L; n-octanol: 0.24 g/L-0.37 g/L), and flow rate (45 L/h-115 L/h). All tests were conducted at a pressure of 24.8 MPa and at least 20% excess oxygen.

The pilot-plant results confirmed the validity of first-order reaction models for SCWO of acetic acid previously developed at The University of Texas at Austin using laboratory-scale apparatus. It appeared that hydrogen peroxide and oxygen were equally effective in SCWO of acetic acid when the oxidant and feed were heated separately and then mixed. First-order reaction models for phenol and n-octanol were also obtained. For practical purposes, pseudo-first-order reaction models for these organic compounds can be used in the design and operation of SCWO processes.

INTRODUCTION

SCWO is emerging as an environmentally attractive R&D option for the safe and economical treatment of toxic organic wastewaters and sludges. It has been demonstrated that the SCWO process is capable of destroying undesirable organic constituents, converting carbonaceous and nitrogenous compounds into non-noxious materials, accomplishing this task in a totally enclosed facility, and efficiently recovering hydrothermal energy. With nearly 10 years of R&D, the process is now poised for commercialization [1, 2].

To facilitate the design of commercial SCWO processes, much effort has been made to obtain kinetic data. However, because of high-temperature (400°C-650°C) and high-pressure (25 MPa-35 MPa) requirements for conducting SCWO experiments, most existing kinetic models for simple organic compounds and organic wastewaters were developed from laboratory-scale studies [3]. These models require pilot-plant validation.

Lixiong Li and Earnest F. Gloyna, Environmental and Water Resources Engineering and Separations Research Programs, The University of Texas at Austin, Austin, TX 78758 (Internal Mail Code 77100)
Peishi Chen, Phillip Morris R&D, P.O. Box 26583, Richmond, VA 23261

The selection of proper model compounds is an important aspect of this special kinetic study because the overall reaction rate is controlled by the rate-limiting transformations. Several studies have identified acetic acid as a major rate-limiting intermediate in SCWO and wet air oxidation of organic compounds [3-7]. Therefore, acetic acid was used in this study. Phenol and n-octanol, representing aromatic and long-chain alcoholic compounds, respectively, were studied.

To validate global kinetic models for the SCWO of these three compounds, a series of pilot-scale tests were conducted [8]. A 150-liter/hour (40 GPH) pilot plant with a 6.1-m long concentric-tube reactor was used. Process variables included reaction temperature (385°C-440°C), feed concentration (acetic acid: 4 g/L-21 g/L; phenol: 0.59 g/L-0.77 g/L; n-octanol: 0.24 g/L-0.37 g/L), and flow rate (45 L/h-115 L/h). All tests were conducted at a pressure of 24.8 MPa and at least 20% excess oxygen. Reactor contents were sampled while the pilot plant was operated under continuous, steady-state conditions.

Kinetic models for SCWO of acetic acid based on laboratory-scale systems were compared with pilot-scale test results. The first-order reaction model developed by Lee [4] using hydrogen peroxide as the oxidant matched the pilot-plant model. Other acetic acid models [7, 9, 10] displayed different degrees of variation from Lee's and this work. First-order models for SCWO of phenol and n-octanol were also obtained. Without mixing the oxidant with the feed prior to heating, hydrogen peroxide and oxygen were equally effective in SCWO of acetic acid. For practical purposes, pseudo-first-order reaction models for these organic compounds can be used in the design and operation of SCWO processes.

PILOT-SCALE SCWO FACILITY

Figure 1 shows a process flow diagram of the 150 liter/hour SCWO pilot plant. Major components in the pilot plant included a high-pressure diaphragm pump, two double-pipe heat exchangers, an electric heater, a concentric-tube reactor, and an air-driven oxygen booster. Process pressures and temperatures were monitored and controlled by a computer workstation. The feed flowrate was controlled by manually adjusting the stroke length of the feed pump, and was monitored by the computer.

The key component of this SCWO facility was the specially designed concentric-tube reactor. A schematic of this reactor is shown in Figure 2. Oxygen was injected into the feed stream after the heater and prior to the reactor inlet. The fluid mixture flowed downwards through the core of the inner tube, and flowed upwards through the annular section between the inner and outer tubes. These configurations allowed a counter-current flow of the reacting fluids, and therefore facilitated the establishment of a more nearly isothermal temperature profile across the fluid flow path, as compared to that obtainable under one-pass tubular reactor designs. The reactor was well insulated to minimize heat loss.

The reactor size was determined by assuming a conversion of 99.99% for some typical organic compounds at a temperature of 450°C and a flow rate of 150 liter/hour based on laboratory-scale kinetic data. The length-to-diameter ratio was determined by the fluid velocity requirement needed to suspend the anticipated particulates and Reynolds number required to simulate ideal plug-flow reactor conditions [8]. The dimensions of the outer tube were 8.89-cm outer diameter, 5.08-cm inner diameter, and 6.1-m long. The dimensions of the inner tube were 1.91-cm outer diameter, 0.17-cm wall thickness, and 6.1-m long. Thermocouples were installed at the reactor inlet, bottom, and outlet, respectively. Similarly provisions were made to collect samples at the reactor inlet, bottom, and outlet at continous and steady-state operation conditions.

PROCEDURES

Pilot-plant operation and analytical measurement procedures are summarized below.

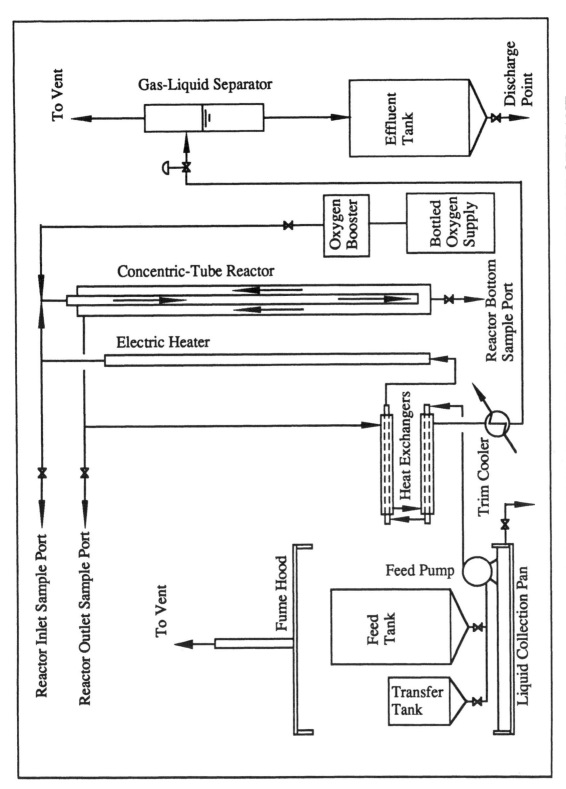

FIGURE 1. PROCESS FLOW DIAGRAM OF 150 LITER/HOUR SCWO PILOT PLANT

221

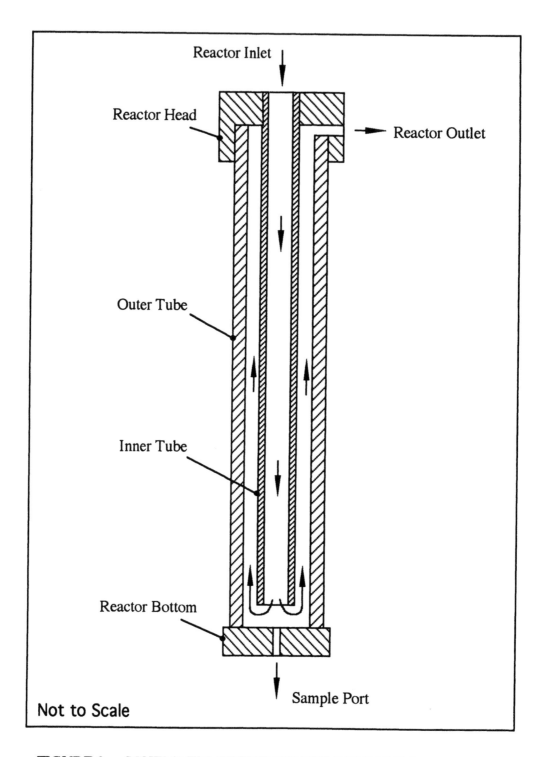

Reactor Inlet

Reactor Head

Reactor Outlet

Outer Tube

Inner Tube

Reactor Bottom

Sample Port

Not to Scale

FIGURE 2. SCHEMATIC OF THE CONCENTRIC-TUBE REACTOR

PILOT-PLANT OPERATION

The pilot-plant operation typically involved (a) feed stock preparation; (b) system conditioning; (c) feed and oxygen introduction; (d) sample collection; and (e) reactor cooling and cleaning.

Organic feed solutions were prepared using deionized water. The feed stock was stirred continuously during each test.

Prior to an actual test, deionized water was pumped through the reactor system at a rate of about 90 liter/hour. During this reactor conditioning phase, the system was pressurized to 24.8 ± 0.7 MPa. After the pressure stabilized, the heat was applied electrically. The temperature setpoint was raised to the desired level in increments of 50°C. All parameters, such as temperature, pressure, and flowrate at various locations of the system, were monitored continuously. Data were collected every five minutes.

The feed pump was switched from water to feed after the desired reactor conditions were established. After this switch occurred, oxygen (medical grade) was introduced at the reactor inlet. About 15 minutes or at least two times the reactor residence time were required for the pilot-plant to reach a steady-state. Then samples were collected at the reactor inlet, outlet, and bottom. After a series of tests were completed, the heater was turned off, and deionized water was pumped into the reactor system. As the reactor was being flushed, the flow of oxygen continued for an additional 15 minutes. This washing sequence continued for about 30 minutes.

ION CHROMATOGRAPH (IC)

An ion chromatograph (Dionex System 14) equipped with an Ion-Pac column (AS-3), Ion-Pac Guard pre-column, and a conductivity meter was used to determine acetate ion concentration. A mixture of sodium bicarbonate (3 mM) and sodium carbonate (2.4 mM) was used as the eluent for the IC operation. Ammonium acetate was used to prepare the standard solutions in conenctrations ranging from 1 mg/L to 1000 mg/L. All test samples were filtered prior to injection.

GAS CHROMATOGRAPH (GC)

A gas chromatograph (Hewlett Packard Model 5890A) was used to determine the concentration of phenol and n-octanol. A fused silica capillary column (Supelco SPB-5) was used to separate the target compounds. The lower detection limits of the flame ionization detector for phenol and n-octanol were 0.5 mg/L and 1 mg/L, respectively. External standards were used for calibration. An average deviation of less than 5% was consistently achieved for most injection responses.

RESULTS AND DISCUSSION

Organic conversions for acetic acid, phenol, and n-octanol were measured as a function of temperature and reactor residence time for a given feed concentration and oxygen-to-feed ratio. The conversion was defined as the amount of the starting compound reacted divided by the amount of the starting compound available at the reactor inlet (= 1-$[C_A]$@effluent/$[C_A]$@inffluent). The reactor bottom temperature was used as the independent variable. Typical variations among the reactor inlet, bottom, and outlet temperature were within ± 3°C. Reactor residence times in the tube and annular sections were calculated using the density of water at the average temperature, flow rate, and volume of each section. The average temperature for the tube section was the arithmetic mean of the reactor inlet and reactor bottom temperatures, and the average temperature for the annular section was the arithmetic mean of the reactor bottom and reactor outlet temperatures. All tests were conducted at greater than 20% excess stoichiometric oxygen demand based on a complete organic conversion. These organic conversion data at various temperatures and reactor residence times were fitted with several reaction rate models based on the rate equation:

$$\text{Rate} = k^{\circ} \exp(-E/RT) [C_A]^m [O_2]^n$$

where E is activation energy (J/gmol); T is temperature (K); R is the gas constant (8.134 J/gmol•K); $[C_A]$ and $[O_2]$ are concentrations of the organic parent compound A and oxygen, respectively; and k° is the pre-exponential factor (the unit depends on the reaction order m and n, for first-order reactions, the unit for k° is sec^{-1}).

However, pseudo-first-order reaction models (first- and zero-order with respect to the organic compound and oxygen, respectively) produced the "best-fit." Both test data and calculated results are given in Table I. Based on these data, Arrhenius plots for acetic acid, phenol, n-octanol were derived as shown in Figures 3 to 5, respectively. It should be noted that pseudo-first-order kinetic models have been determined via the nonlinear regression for a number of simple compounds, such as acetic acid, ethanol, and methanol [9, 12-13].

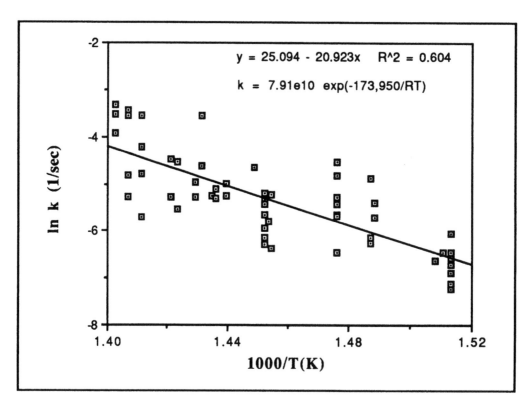

FIGURE 3. ARRHENIUS PLOT FOR SCWO OF ACETIC ACID.

Furthermore, the kinetic parameters for these model compounds obtained from this study and reported by others are summarized in Table II. The activation energy for acetic acid obtained from 57 pilot-plant data points was 174 kJ/mol. Wightman [9] reported, based on seven data points, that SCWO of acetic acid followed a second-order reaction (first-order with respect to acetic acid and oxygen), and the activation energy was 231 kJ/mol. When these seven data points were fitted with a pseudo-first-order reaction model, the activation energy became 172.7 kJ/mol. Kinetic data for SCWO of acetic acid using hydrogen peroxide (H_2O_2) as the oxidant were also available [4, 10]. Lee's study using a coiled tube reactor resulted in a slightly lower activation energy, 167.1 kJ/mol [4] as compared to the pilot-plant results. Wilmanns used a small, concentric-tube reactor [10] and he reported somewhat different pre-exponential factor and activation energy values as compared to Lee's

TABLE I – PILOT-PLANT TEST AND CALCULATED DATA

Organic Compound	Run Number	Influent Conc. (g/L)	Effluent Conc. (g/L)	Conversion X	Bottom Temperature (K)	Residence Time (s)	Rate Constant (1/s)
HAc	Hac-1	20.25	16.1	0.205	688.6	50.92	.00450
	Hac-3	19.5	16.5	0.154	688.6	63.16	.00264
	Hac-4	19.5	18.9	0.031	688.6	6.47	.00483
	Hac-5	18.75	16.0	0.147	688.6	84.1	.00189
	Hac-6	18.75	18.4	0.019	688.6	8.74	.00216
	Hac-7	19.5	12.75	0.346	688.6	121.1	.00351
	Hac-8	19.5	18.62	0.045	688.6	13.1	.00350
	Hac-9	19.5	15.58	0.201	677.4	142.5	.00157
	Hac-11	20.5	14.55	0.29	677.4	96.72	.00354
	Hac-12	20.5	18.9	0.276	677.4	9.91	.0082
	Hac-13	19.9	14.4	0.29	677.4	73.44	.0044
	Hac-14	19.9	19.4	0.025	677.4	7.4	.00344
	Hac-15	20.25	15.0	0.259	677.4	59.15	.00507
	Hac-16	20.25	19.0	0.062	677.4	5.88	.0108
	Hac-17	20.5	6.3	0.693	710.8	41.57	.0284
	Hac-18	20.5	17.75	0.134	710.8	4.42	.0326
	Hac-19	19.62	17.5	0.108	660.8	92.92	.00123
	Hac-21	21.12	17.84	0.156	660.8	114.6	.00147
	Hac-22	21.12	20.75	0.018	660.8	11.4	.00157
	Hac-23	20.5	16.56	0.192	660.8	150.3	.00142
	Hac-24	20.5	20.25	0.012	660.8	15.2	.00081
	Hac-25	20.25	15.6	0.229	696.9	49.22	.00529
	Hac-27	20.25	12.25	0.395	696.3	81.66	.00616
	Hac-28	20.25	19.37	0.043	696.3	8.74	.00505
	Hac-29	20.25	11.49	0.433	694.7	82.77	.00685
	Hac-30	20.25	19.33	0.045	694.7	8.82	.00525
	Hac-31	4.937	4.375	0.114	660.8	165.9	.00073
	Hac-32	4.937	4.75	0.038	660.8	16.38	.00236
	Hac-33	5.062	4.375	0.136	663	109.8	.00133
	Hac-35	5.062	4.375	0.136	661.9	91.77	.00159
	Hac-38	4.313	4.187	0.029	671.9	6.57	.00451
	Hac-40	4.375	4.25	0.029	671.9	8.63	.00336
	Hac-41	4.625	3.25	0.297	710.8	67.84	.0052
	Hac-42	4.625	4.375	0.054	710.8	6.83	.00814
	Hac-43	4.25	0.325	0.924	708.6	89.55	.0287
	Hac-44	4.25	4.125	0.029	708.6	9.08	.00329
	Hac-45	5.062	2.625	0.481	708.6	78.6	.00835
	Hac-46	5.062	4.5	0.111	708.6	7.93	.0148
	Hac-47	4.75	2.25	0.526	699.7	105.6	.00707
	Hac-48	4.75	4.25	0.053	699.7	10.69	.00506
	Hac-49	4.75	3.5	0.263	702.4	76.85	.00397
	Hac-50	4.75	4.375	0.079	702.4	7.71	.0107
	Hac-51	4.375	2.25	0.486	703.6	57.45	.0116
	Hac-52	4.375	4.25	0.029	703.6	5.71	.00508
	Hac-53	4.75	2.875	0.395	698.6	50.23	.01
	Hac-54	4.75	4.125	0.132	698.6	4.95	.0285
	Hac-55	4.625	2.75	0.405	690.2	53.86	.00965
	Hac-57	5.062	3.5	0.309	688.6	66.12	.00558
	Hac-59	4.062	3.5	0.138	687.4	87.44	.0017
	Hac-60	4.062	3.875	0.046	687.4	8.7	.00542

TABLE I – PILOT-PLANT TEST AND CALCULATED DATA (CONTINUED)

Organic Compound	Run Number	Influent Conc. (g/L)	Effluent Conc. (g/L)	Conversion X	Bottom Temperature (K)	Residence Time (s)	Rate Constant (1/s)
	Hac-61	4.687	3.312	0.293	688	112.7	.00308
	Hac-64	4.625	4.5	0.027	672.4	14.07	.00195
	Hac-65	4.75	3.75	0.211	672.4	109.6	.00216
	Hac-66	4.75	4.375	0.079	672.4	10.8	.00761
	Hac-67	4.705	1.112	0.764	713	49.23	.0293
	Hac-68	4.705	4.275	0.091	713	4.9	.0196
	Hac-69	4.705	.812	0.827	713	49.23	.0357
Octanol	Oct-1	.318	.0114	0.964	715.2	4.83	.689
	Oct-2	.331	.0085	0.974	710.8	6.44	.568
	Oct-3	.366	.0068	0.981	710.8	8.4	.474
	Oct-4	.293	.0055	0.981	710.8	10.95	.363
	Oct-5	.330	.0167	0.949	699.7	9.37	.318
	Oct-6	.345	.0262	0.924	699.7	6.95	.371
	Oct-7	.355	.0634	0.821	699.7	5.48	.314
	Oct-8	.324	.0783	0.758	688.6	5.99	.237
	Oct-9	.232	.0248	0.893	713.6	5.38	.416
	Oct-10	.259	.0167	0.935	713.6	6.33	.433
	Oct-11	.268	.0439	0.836	688.6	7.6	.238
	Oct-12	.277	.0301	0.891	688.6	9.82	.226
	Oct-13	.257	.0132	0.949	688.6	12.9	.230
	Oct-14	.261.5	.0204	0.922	677.4	14.72	.173
	Oct-15	.240	.0285	0.881	677.4	11.37	.187
	Oct-16	.265	.0569	0.785	677.4	8.5	.181
Phenol	Ph-1	.6556	.0856	0.869	715.2	48.66	.0418
	Ph-2	.6556	.3496	0.467	715.2	4.83	.13
	Ph-3	.6485	.0523	0.919	710.8	64.4	.0391
	Ph-5	.7402	.0199	0.973	710.8	83.86	.0431
	Ph-6	.7402	.2356	0.682	710.8	8.40	.1363
	Ph-7	.6503	.0049	0.993	710.8	108.65	.045
	Ph-8	.6503	.1745	0.732	710.8	10.95	.12
	Ph-9	.7053	.0177	0.975	699.7	93.25	.0395
	Ph-10	.7053	.3643	0.483	699.7	9.37	.0705
	Ph-11	.7004	.0146	0.979	699.7	69.4	.0558
	Ph-12	.7004	.387	0.447	699.7	6.95	.0854
	Ph-14	.7724	.5166	0.331	699.7	5.48	.0734
	Ph-16	.6732	.5233	0.223	688.6	5.99	.0421
	Ph-18	.587	.400	0.318	713.6	5.38	.0711
	Ph-19	.616	.045	0.927	713.6	63.39	.0413
	Ph-20	.616	.3459	0.438	713.6	6.33	.0912
	Ph-21	.616	.0686	0.889	688.6	76.11	.0288
	Ph-22	.616	.3788	0.385	688.6	7.6	.064
	Ph-23	.638	.0301	0.953	688.6	97.73	.0312
	Ph-24	.638	.3635	0.430	688.6	9.82	.0573
	Ph-25	.658	.0030	0.995	688.6	127.9	.0421
	Ph-26	.658	.3122	0.526	688.6	12.9	.0578
	Ph-27	.665	.001	0.999	677.4	145.8	.0444
	Ph-28	.665	.3318	0.501	677.4	14.72	.0472
	Ph-29	.603	.0135	0.9776	677.4	113.1	.0336
	Ph-30	.603	.3324	0.4488	677.4	11.37	.0524
	Ph-31	.636	.0404	0.9365	677.4	85.0	.0324

and this work. A wet air oxidation kinetic model reported involving first-order reaction with respect to acetic acid and 0.37-order with respect to oxygen [7]. In Figure 6, Arrhenius relationships for these reported data are compared with the pilot-plant data on a pseudo-first-order reaction basis. The comparable kinetic results for acetic acid exposed to oxygen and hydrogen peroxide suggest that reaction pathways were similar in both cases once fluid flow was fully developed in the reaction zone and possibly some hydrogen peroxide may have decomposed to oxygen and water during the heating stage.

As shown in Figure 4, the pilot-plant data for phenol were more scattered when treated with first-order reaction models ($r^2 = 0.231$), as compared to those for acetic acid and n-octanol. The activation energy, based on 27 data points, was 60.8 kJ/mol. Reported kinetic models for SCWO of phenol based on laboratory-scale studies have included oxygen concentration term [9, 11] and water concentration term [11]. Possibly phenol reaction pathways in the SCWO environment were substantially complex, and the role of oxygen and water may require explicit consideration as expressed in the reported kinetic models [9, 11]. In addition, this mechanism may be more sensitive to process conditions. For example, experiments reported by Thornton and Savage [11] were conducted near the critical point of water ($0.89 \leq T_r \leq 1.07$; $0.86 \leq P_r \leq 1.27$). Many properties of water change drastically across the critical point, and these changes may have significant effects on the reaction mechanism. Nevertheless, the activation energy values for SCWO of phenol in all three cases were similar, ranging from 52 kJ/mol to 64 kJ/mol.

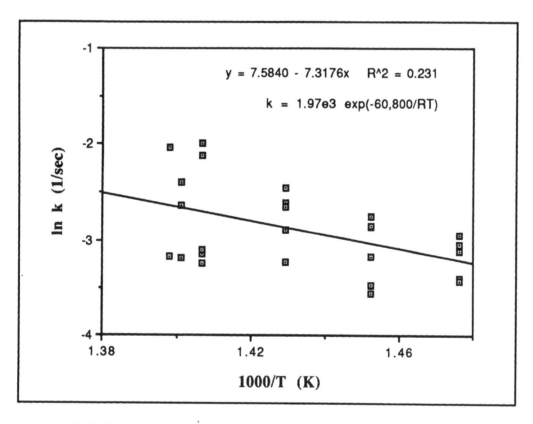

FIGURE 4. ARRHENIUS PLOT FOR SCWO OF PHENOL.

The kinetic parameters for n-octanol were based on 16 data points. As shown in Figure 5, a much better first-order fitted was obtained for n-octanol as compared to those discussed earlier for acetic acid and phenol. The activation energy value was 115.5 kJ/mol. No literature data were available for comparison. As an additional reference, kinetic parameters for methanol and ethanol are given in Table II.

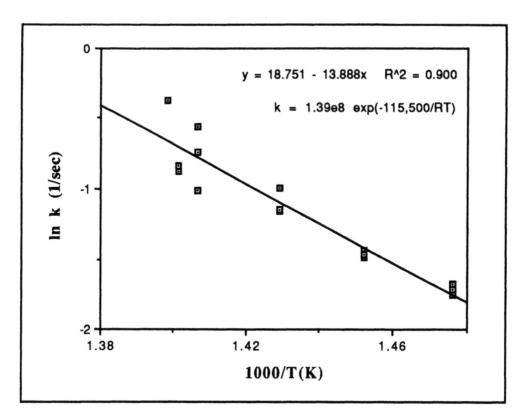

$$y = 18.751 - 13.888x \quad R^2 = 0.900$$

$$k = 1.39e8 \; \exp(-115{,}500/RT)$$

FIGURE 5. ARRHENIUS PLOT FOR SCWO OF n-OCTANOL.

Table II – KINETIC MODELS FOR SCWO OF SELECTED ORGANIC COMPOUNDS

Compound	Parameters				Temperature (K)	Pressure (MPa)	Feed Conc. (g/L)	Reference
	$k°$	E	m	n				
Acetic Acid	9.82×10^{17}	231	1	1	611-718	40-44	0.525	[9]
Acetic Acid‡	1.57×10^{11}	170	1	0	611-718	40-44	0.525	[9]
Acetic Acid*	2.63×10^{10}	167.1	1	0	673-803	24-35	1.3-3.3	[4]
Acetic Acid*‡	9.23×10^{7}	131	1	0	673-803	24-35	1.0-5.0	[10]
Acetic Acid#	5.60×10^{10}	167.7	1	0.37	543-593	10-20	29-38	[7]
Acetic Acid#‡	4.40×10^{12}	182	1	0	543-593	10-20	29-38	[7]
Acetic Acid	7.91×10^{10}	174	1	0	660-713	25	4.1-20.5	[This work]
Phenol	2.61×10^{5}	64.0	1	1	557-702	30-35	0.1-0.4	[9]
Phenol‡#	6.00×10^{3}	59.0	1	0	557-647	30-35	0.1-0.4	[9]
Phenol†	3.03×10^{2}	51.8	1	0.5	573-693	19-28	0.05-0.33	[11]
Phenol†‡#	8.58×10^{1}	41.7	1	0	573-647	28	0.05-0.33	[11]
Phenol	1.97×10^{3}	60.8	1	0	677-715	25	0.59-0.74	[This work]
Octanol	1.39×10^{8}	115.5	1	0	677-715	25	0.24-0.37	[This work]
Ethanol	6.46×10^{21}	340	1	0	755-814	24	0.03-0.036	[12]
Methanol	1.58×10^{26}	408	1	0	723-803	25	0.041-0.18	[13]

*	H_2O_2 was used as the oxidant, while in all other cases O_2 was used.
#	Subcritical water oxidation.
‡	Reported data were re-fitted with pseudo-first-order models.
†	Rate = $k° \exp(-E/RT)[Phenol]^{1.0}[O_2]^{0.5}[H_2O]^{0.7}$, the unit of $k°$ is $(gmol/L)^{-1.2} \cdot s^{-1}$.

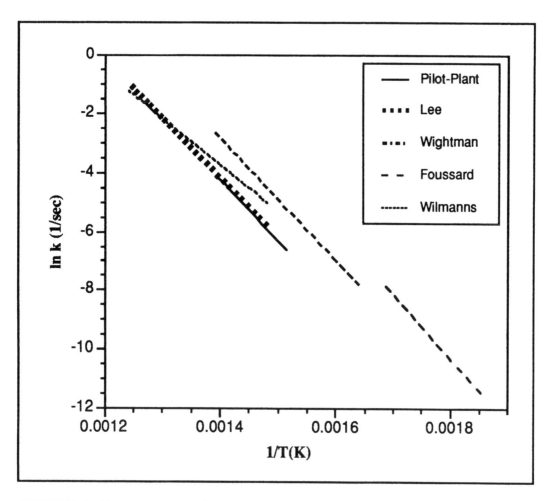

FIGURE 6. COMPARISION OF PSEUDO-FIRST-ORDER KINETIC MODELS
FOR SUPERCRITICAL WATER OXIDATION OF ACETIC ACID.

CONCLUSIONS

The first-order reaction model for SCWO of acetic acid derived from this pilot-plant study was similar to that obtained by Lee [4] using hydrogen peroxide as the oxidant. In both cases, the feed and oxidant were heated separately and then mixed prior to entering the reactor. The comparable kinetic models for SCWO of acetic acid using either oxygen or hydrogen peroxide suggest that similar reaction pathways may exist in these two cases, and some hydrogen peroxide may have possibly decomposed to form oxygen and water during the heat-up period.

SCWO of n-octanol appeared to fit the first-order kinetic model under the pilot-plant test conditions. However, the pilot-plant data for SCWO of phenol were somewhat scattered when treated with first-order reaction models. The role of oxygen and water in phenol reaction in SCWO envrionment may require explicit consideration as expressed in the kinetic models by Wightman [9] and Thornton [11].

For practical purposes, pseudo-first-order reaction models can be adequate for the design and operation of SCWO waste treatment processes.

ACKNOWLEDGEMENT

Appreciation is extended to Eco Waste Technologies (Austin, Texas) for providing financial and technical assistance in the pilot-plant construction and operation.

REFERENCES

1. McBrayer, R., L. Li, and E. F. Gloyna. 1993. "Research and Development of a Commercial Supercritical Water Oxidation Process," in Proceedings of Eleventh Annual Environmental Management and Technology Conference/International, Advanstar Expositions, pp. 90-111.

2. Rosasco, G. J. (Ed.). September 1992. Proceedings: Workshop on Federal Programs Involving Supercritical Water Oxidation, NISTIR 4920, National Institute of Standards and Technology.

3. Li, L., P. Chen, and E. F. Gloyna. 1991. "Generalized Kinetic Model for Wet Oxidation of Organic Compounds." American Institute of Chemical Engineers Journal, 37 (11), 1687-1697.

4. Lee, D. S.. August 1990. "Supercritical Water Oxidation of Acetamide and Acetic Acid," Ph.D. Dissertation, Department of Civil Engineering, The University of Texas at Austin.

5. Shanableh, A. M.. December 1990. "Subcritical and Supercritical Water Oxidation of Industrial, Excess Activated Sludge," Ph.D. Dissertation, Department of Civil Engineering, The University of Texas at Austin.

6. Taylor, J.E., and J.C. Weygandt. 1974. "A Kinetic Study of High Pressure Aqueous Oxidations of Organic Compounds Using Elemental Oxygen." Canadian Journal of Chemistry, 52: 1925-1933.

7. Foussard, J.-N.. 1983. "Study on Aqueous Phase Oxidation at Elevated Temperatures and Pressures," Ph.D. Dissertation, National Institute of Applied Science, Toulouse, France.

8. Gloyna, E. F. and L. Li. August 1991. Characterization of Shallow, Subsurface Supercritical Water Oxidation Reactors," Year One Report, Department of Civil Engineering, The University of Texas at Austin.

9. Wightman, T. J.. June 1981. "Studies in Supercritical Wet Air Oxidation," MS Thesis, Department of Chemical Engineering, University of California, Berkeley.

10. Wilmanns, E. G.. December 1990. "Supercritical Water Oxidation of Volatile Acids," MS Thesis, Department of Civil Engineering, The University of Texas at Austin.

11. Thornton, T. D. and P. E. Savage, 1992. "Kinetics of Phenol Oxidation in Supercritical Water." American Institute of Chemical Engineers Journal, 38(3): 321-327.

12. Helling, R. K.. 1986. "Oxidation Kinetics of Simple Compounds in Supercritical Water: Carbon Monoxide, Ammonia and Ethanol," Ph.D. Dissertation, Department of Chemical Engineering, Massachusetts Institute of Technology, Cambridge.

13. Tester, J. W., P. A. Webley, and H. R. Holgate, 1993. "Revised Global Kinetic Measurements of Methanol Oxidation in Supercritical Water." Industrial and Engineering Chemistry Research, 32(1): 236-239.

DANIEL CAMPOS
K. L. McDOWELL
C. A. TOLMAN

Thermodynamic Analysis of Wet Oxidation

ABSTRACT

The wet air oxidation (WAO) concept has been around for decades. In recent years, interest in WAO has increased as an option for the treatment of aqueous organics that are too concentrated for biotreatment or too dilute for incineration. The purpose of this study is to determine the ranges of operating conditions where wet oxidation is applicable in the mineralization of aqueous organics. This was done through a rigorous thermodynamic analysis of the process as is described in academic and vendor literature (namely, adiabatic liquid phase oxidation of organic carbon to CO_2 and water). The analysis involved mass/energy balances and liquid-vapor/equilibrium calculations on a simulated wastewater containing phenol. The calculations, done with the ASPEN+ simulation software, show that the range of applicability (T, P, concentration ranges) of the wet oxidation process run adiabatically is very limited if the amount of water evaporated in the reactor is constrained. The constraint on water evaporation is needed for waste streams containing inert compounds with limited solubility. This limitation could be relaxed by running the oxidation reaction at constant temperature if this were economically feasible or by drastically reducing the reaction temperature if the reaction kinetics would allow reasonable reaction rates.

WAO DESCRIPTION AND CAPABILITIES

Wet air oxidation (see Fig. 1) is the liquid phase oxidation of complex organics and/or oxidizable inorganic components by air or oxygen in the presence of <u>liquid water</u>. Elevated temperatures (400-600°F or 204-316°C) are required to achieve useful reaction rates. To control the rate of vaporization and <u>maintain a liquid phase</u>, elevated pressures, ranging from 300 to over 3000 pounds per square inch, are required. The process is operated continuously with the wastewater and gaseous source of oxygen (usually air) streams fed into the reactor simultaneously. The oxygen in the air reacts with the organic matter in the wastewater to produce mainly carbon dioxide, water and simpler organics like acetic acid.

Controlling the amount of water evaporated in the reactor becomes important when the waste stream contains inorganic compounds with limited solubility in water. When too

Daniel Campos, K. L. McDowell and C. A. Tolman, E. I. du Pont de Nemours and Co., Route 141, Bldg. 262, Rm. 312, Wilmington, Delaware 19880, USA much water evaporates, some of the salt may precipitate and foul the reactor and heat exchangers.

Due to the exothermic nature of the reaction, heat is produced which raises the temperature of the reactor contents. The effluent from the wet oxidation reactor is heat exchanged with the incoming waste. If the COD reduction is higher than about 15,000 ppm, no auxiliary heat is required. The maximum requirement for auxiliary heat is between 500 and 600 BTU/gal [1].

The Zimpro process reports the chemical oxygen demand (COD) removal to typically be about 75% to 90% with residence times ranging from 15 min to 120 min. The operating pressure in the Zimpro process is maintained well above the saturation pressure corresponding to the operating temperature so that the reaction is carried out in the liquid phase [2].

Kenox reports that it can treat streams with COD content in the range 1-35%. The typical operating ranges are 200-250 $^{\circ}$C and 500-800 psi, while the COD reductions are between 40 and 90+% [3].

Two Commercial scale wet oxidation treatments were conducted at the Casmalia Resources facility in Santa Barbara County, California using Zimpro technology. One process involved the oxidation of 5.3 GPM of Gulf oil spent caustic wastewater containing phenols at 268 $^{\circ}$C and 1610 psig, nominal residence time of 113 min, 190 SCFM of air, and residual oxygen concentration of 3.7%. The influent COD was 108 g/L and the effluent COD was 12 g/L, which gives a reduction of 89 %. The influent DOC was 20.8 g/L and the effluent DOC was 0.7 g/L, which gives a reduction of 97%. The influent phenol concentration was 15.5 g/L and the effluent was 0.04 g/L, which gives a reduction of 99.8% [1, 4]

The second Zimpro commercial-scale WAO process involved the treatment of 7.5 GPM of cyanide wastewater at 257 $^{\circ}$C and 1220 psig, nominal residence time of 80 min, 190 SCFM of air, and residual oxygen concentration of 7.1%. The influent COD was 37.4 g/L and the effluent COD was 4.2 g/L, which gives a reduction of 88.8%. The influent DOC was 14.71 g/L and the effluent DOC was 1.71 g/L, which gives a reduction of 88.4% [1, 4].

WAO KEY FEATURES

From the open literature, one can surmise the following features of WAO:

- The reactor is run adiabatically (i.e., no heat is added or removed)

- Most of the water is maintained in the liquid phase

- It can treat streams with high organic content

- It mineralizes much of the organic carbon, i.e., it converts it to CO_2

- It can remove about 90% of the organic carbon

- It is autothermal if the COD reduction is at least about 1.5%

OBJECTIVE OF THIS STUDY

The purpose of this study has been to determine the ranges of temperature, pressure, and concentration of organics at which WAO mineralizes 90% of the organic carbon in a phenol stream while maintaining most of the water in the liquid phase. Another objective was to determine the conditions at which the process is autothermal. This was done through a rigorous thermodynamic analysis, which involved mass and energy balances and liquid-vapor equilibrium calculations. The study of the reaction kinetics is outside the scope of this paper.

MODEL AND ASSUMPTIONS

An abbreviated form for the thermodynamic model around any given component of the process (e.g., reactor) can be written as follows:

Mass balance

$$\sum[(m_i)_{in} - (m_i)_{out}] = 0 \qquad (1)$$

Energy balance

$$\sum[(m_i h_i)_{in} - (m_i h_i)_{out}] = Q \qquad (2)$$

Vapor-liquid equilibrium

$$y_i = K_i x_i \qquad (3)$$

where:

m_i = mass flowrate of component i
h_i = enthalpy of component i
Q = heat flowrate
y_i = concentration of component i in the vapor phase
x_i = concentration of component i in the liquid phase
K_i = vapor-liquid equilibrium constant of component i

The fraction of water that evaporates in the waste/air mixture at the reactor outlet is a function of the temperature and pressure. The outlet temperature is a function of the inlet temperature, heat of reaction and the reactor mode of operation (e.g., adiabatic). For the adiabatic case, the calculation involves trial and error. The outlet temperature must be guessed, which allows one to calculate the compositions of the liquid and vapor phases (assuming vapor/liquid equilibrium). The correct temperature is that which satisfies the mass and energy balances and the liquid-vapor equilibrium equations.

An ASPEN+ model was set up to do mass/energy balances and liquid-vapor equilibrium calculations throughout the flow-sheet shown in Fig. 1. The discussion in this report focuses on the effects of the operating conditions on the fraction of water that evaporates in the reactor and the auxiliary heat requirements of the process.

The model key assumptions are:

1. The waster stream flowrate is 20 gal/min and is contaminated with a specified weight % of phenol

2. All process components operate at steady-state

3. The liquid and vapor phases in each stream are at equilibrium

4. The theoretically required (stoichiometric) amount of air is supplied to the process, i.e., just enough oxygen is supplied to convert all the phenol present to CO_2 and water

5. The only reaction occurring is $C_6H_5OH + 7 O_2 \rightarrow 6 CO_2 + 3H_2O$[1]

6. At the given conditions the reaction is assumed to proceed to 90% conversion

7. The heat exchanger has an area 500 ft^2 and the overall heat transfer coefficient (U) is 60 BTU/hr-ft^2-^{o}F

8. Whenever possible the heat exchanger transfers just enough heat to preheat the reactor influent to the specified reactor inlet temperature. Any deficit is made up by the auxiliary heater.

9. The thermodynamic properties are obtained from the BWR-Lee-Starling equation of state, which is particularly suitable in the high temperature and high pressure regions explored here

---

[1] One might argue that there are many possible byproducts from the oxidation reaction and the simulation results would depend on the reaction pathway chosen. However, for a given COD reduction, the results of the simulation are nearly independent of the reaction pathway and byproducts This is because the heat of reaction is the determining factor, and is practically independent of the byproducts as long as the COD reduction is fixed.

RESULTS AND DISCUSSION

A series of adiabatic (Q=0 in the reactor) wet oxidation simulation runs was done in which the reactor inlet temperature was varied between 150 and 300 °C, the pressure was varied between 1000 and 3000 psi, and the phenol concentration was varied between 1 and 5% by weight. All runs were done with the theoretically required air and with a phenol conversion of 90%, assuming steady state. The results for the fraction of water evaporated at the reactor outlet are displayed in Figure 2.

As expected, at a given reactor inlet temperature and phenol concentration, increasing the pressure decreases the fraction of water evaporated in the reactor. Also, at a given pressure, increasing the inlet temperature increases the fraction.

Let's assume that the feasibility of the WAO process constrains the fraction of water evaporated to no more than 20%, and the inlet reactor temperature[2] to 250 °C. One can conclude from Fig. 2 that adiabatic WAO under these conditions is not feasible for streams containing more than about 1.5% phenol by weight even when the reactor is at 3000 psi.

Clearly, adiabatic WAO is feasible for treating a 1% phenol stream at 250 °C and 2500 psi (see Fig. 2), but this process requires about 1180 BTU/gal of auxiliary heating, i.e., it is not autothermal (see Fig. 3).

Figure 3 shows the auxiliary heat requirements for an adiabatic WAO process run at 90% mineralization at reactor inlet temperatures of 200 and 250 °C. It can be seen that where WAO is applicable (subject to the constraint of less than 20% evaporation), it generally requires auxiliary heat. Increasing the pressure reduces the auxiliary heat requirement only slightly. For instance, WAO at 250 °C is feasible on a 1% phenol sol. at 2500 psi or higher. If the pressure is increased to 3000 psi, the auxiliary heat requirement is lowered from 1180 to only 1080 BTU/gal.

Decreasing the reactor inlet temperature increases the feasible range of adiabatic WAO. For instance at reactor inlet temperature of 200 °C a 2% phenol stream can be mineralized by 90% at 2000 psi or higher. This assumes that the reaction occurs at a reasonable rate at the chosen initial reaction temperature, which may not be necessarily the case unless a suitable catalyst is used.

Controlling the temperature throughout the reactor is an effective way of limiting the fraction of water that evaporates. A limiting case is when the reactor is run isothermally (see Figure 4).

For example, a WAO treatment process on 5% phenol with 90% conversion and less than 20% water evaporation is possible if run isothermally at 250 °C and a pressure of at least 1900 psi (see Fig. 4). This requires, however, that the reactor be cooled at the rate of about 5200 BTU/gal. Also, because the reactor effluent is cooler than if it had been run adiabatically, about 2000 BTU/gal of auxiliary heat is required by the isothermal process, i.e., the process is not autothermal.

The results on water evaporation and auxiliary heat requirement obtained here apply to a water stream containing phenol. In general, the amount of water evaporated in adiabatic WAO is proportional to the heat of reaction, which is proportional to the standard heat of combustion and the amount of carbon mineralized. Table I shows the normalized heat

---

[2]A previous study [5] shows that more than 80% of the total organic carbon in a phenol sol. can be destroyed within 30 min at a temperature greater than 232 °C.

of combustion (per unit mass of organic carbon) produced in the mineralization of various organics. This value is within -9% and +42% of that for phenol. Therefore, for equivalent mass of carbon mineralized and process conditions, the amount of water evaporated in the WAO of a stream involving almost any other compound than phenol would be at least the calculated value for a phenol stream.

Table I. Normalized heats of combustion for various compounds. The heat of combustion was calculated from the standard heat of reaction in the oxidation reaction [organic + a O_2 → b CO_2 + c H_2O]. The heat of combustion was normalized by dividing the heat of combustion (Kcal/gmole) by the carbon content in the compound (g/gmole).

Compound	Normalized heat of combustion (Kcal/g)
Benzene	10.9
Phenol	10.2
Biphenyl	10.4
Benzyl alcohol	10.6
Benzaldeheyde	10.0
Methanol	14.5
Butanol	13.3
Hexanol	13.2
Octanol	13.2
Decanol	13.1
Cyclohexanol	12.4
Diethyleneglycol	10.6
Acetic acid	8.7
Adipic acid	9.3
Acetone	11.9
1,4 Dioxane	11.8
Tetrahydrofuran	12.5

CONCLUSIONS

1. Adiabatic liquid-phase wet oxidation to a high conversion (90% mineralization) is feasible only in fairly dilute streams, for instance, those containing about 1% TOC or less. If this requires a reactor inlet temperature of 250 ºC, the pressure must be at least 2500 psi to prevent more than 20% water evaporation.

2. The limitations on adiabatic wet oxidation could be reduced somewhat if the reactor inlet temperature could be reduced, namely if the reaction still occurred at a reasonable rate at the lower temperature. For instance liquid wet oxidation of 1% phenol to 90% mineralization would be feasible at 1000 psi if the reactor inlet temperature were 200 ºC.

3. If adiabatic WAO oxidation could be run at reasonable rates and high conversions at reactor inlet temperatures of 150 ºC or less this would extend the WAO applicability to more concentrated solutions. This would also simplify the equipment and materials of construction. For instance, adiabatic wet oxidation of a 3% phenol solution to 90% mineralization is thermodinamically possible with less than 20% water evaporation if the inlet temperature is 150 ºC and the pressure is at least 1500 psi.

4. In the range where adiabatic wet oxidation is feasible with less than 20% water evaporation the process is usually not autothermal, but requires auxiliary heat. For instance, 90% mineralization of 1% phenol is feasible at 250 ºC and 2500 psi, but requires about 1180 BTU/gal in auxiliary heat.

5. The limitations of adiabatic wet oxidation are not significantly removed with the use of pure oxygen as opposed to air. Also, increasing the excess air does not reduce the amount of water evaporated, but it actually increases it slightly.

6. One way to extend the applicability of wet oxidation to more concentrated organic streams is to run the reactor isothermally, namely, with reactor cooling.. For instance, 90% mineralization of 5% phenol is feasible in an isothermal reactor at 250 ºC and 2000 psi. The isothermal WAO reactor requires cooling at the rate of about 5200 BTU/gal. This would of course complicate the wet oxidation process and increase its capital cost.

7. The high TOC reductions in very concentrated wastes observed in other studies [1, 4] under adiabatic WAO conditions may be explained by the fact that the TOC may have been reduced by a mechanism different than mineralization. For instance, partial oxidation reactions can lead to compounds that may polymerize and then precipitate. Another possible explanation is that some TOC may be removed by air stripping.

REFERENCES

1. Hazardous Waste Treatment Technologies: Wet Oxidation, A. P. Jackman and R. L. Powell, Noyes Publications, Park Ridge, NJ, (1991)

2. Generalized Kinetic Model for Wet Oxidation of Organic Compounds, Lixiong Li, Peishi Chen, and Earnest F. Gloyna. AIChE Journal, November 1991, Vol.37, No.11.

3. Kenox WAO process advertisement brochure

4. Standard Handbook of Hazardous Waste Treatment and Disposal, Harry M. Freeman (ed.), McGrawHill, New York 1989. Section 8.6, WET OXIDATION, W.M. Copa, Vice President, Technical Services Zimpro Inc., W.B. Gitchel.

5. Fate of specific pollutants during wet air oxidation and ozonation, Baillod, C.R., Faith, B.M., and Masi, O. , Environ. Prog., 1(3), 217 (1982)

FIG. 1   WET AIR OXIDATION PROCESS

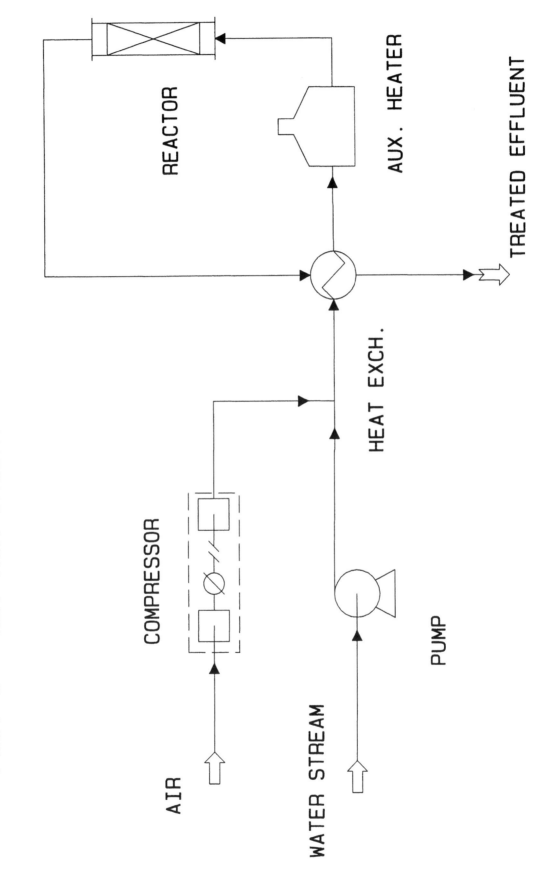

239

Figure 2. Fraction of water evaporated at the reactor outlet when a phenol stream is adiabatically wet oxidized to 90% mineralization

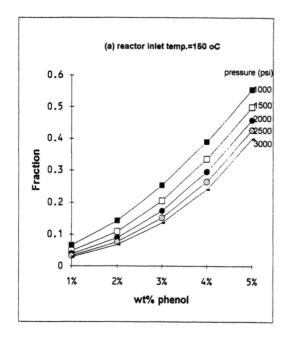

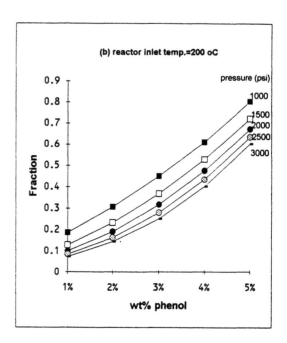

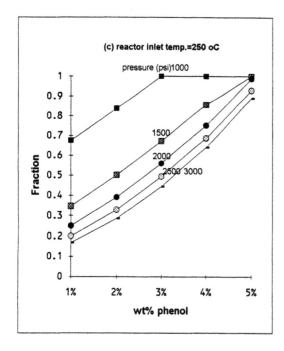

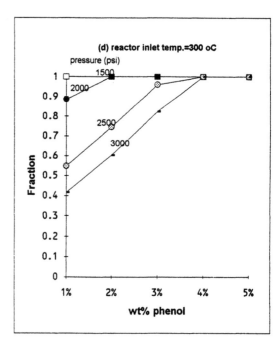

Fig. 3   Auxiliary heat requirement by the adiabatic wet air oxidation of phenol contaminated water with the theoretically required air and 90% mineralization

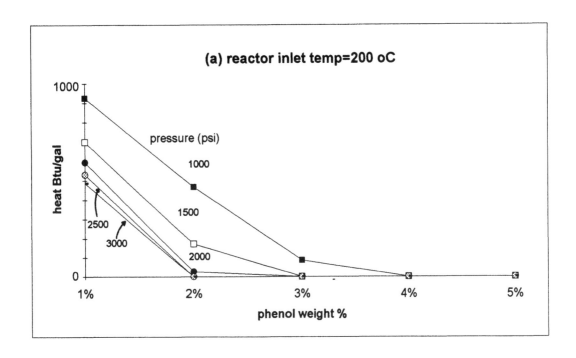

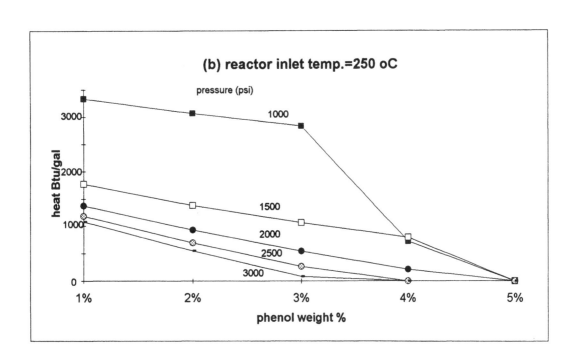

Fig. 4   Fraction of water evaporated at the reactor outlet in the isothermal WAO process at 250 oC on 5%phenol to 90% mineralization

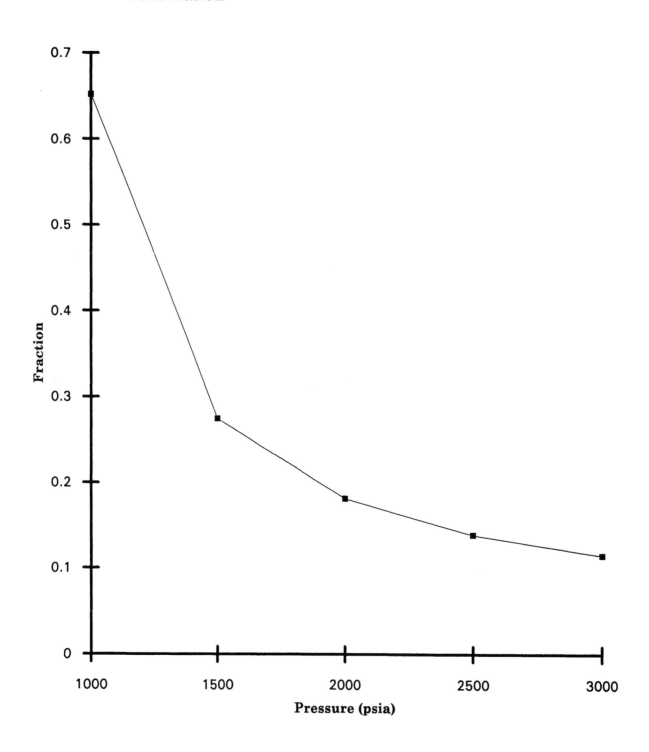

JOSEPH A. MOMONT
WILLIAM M. COPA

The Development and Operation of Full Scale Wet Air Oxidation Units for the Treatment of Industrial Wastewaters

ABSTRACT

Wet air oxidation is a proven technology for effectively treating industrial wastewaters. The wet air oxidation process is used to oxidize organic and some inorganic compounds in an aqueous phase, utilizing elevated pressures (690 to 20700 KPa) and elevated temperatures (120 to 350°C). In a wet air oxidation system, organic compounds are generally oxidized to carbon dioxide and low molecular weight oxygenated organics such as acetic and formic acid. The reduced forms of sulfur, both organic and inorganic, are easily oxidized to sulfate. The oxidation of nitrogenous organic compounds generally yields inorganic ammonia, nitrate or elemental nitrogen. Typically, the by-products in the oxidized effluent are easy to treat biologically. The wet air oxidation process is best fitted for wastewaters which are initially too toxic to be biologically treated and/or too dilute or corrosive to incinerate.

A full scale wet air oxidation system has been supplied for the destruction of para-aminophenol and sulfide present in an industrial wastewater. The wastewater is a by-product from the manufacturing of acetaminophen. Another wet air oxidation system has been provided to a commercial treater of industrial wastewaters. This wet air oxidation system is currently being used to treat a variety of industrial wastes arising from the production of pharmaceuticals, fine chemicals and intermediates, agrichemicals, and phenolic resins.

Results from the development tests, which were used to design wet air oxidation systems, will be discussed in the paper. The systematic experimentation performed in both bench scale batch testing and pilot scale continuous flow testing will be compared to the operating data obtained from the full scale wet air oxidation system. A discussion of the operational availability and acceptability of the wet air oxidation systems is presented.

Joseph A. Momont, Materials/Process Specialist, Dr. William Copa, Vice-President of Technical Services, Zimpro Environmental, Inc. 301 West Military Road, Rothschild Wisconsin, 54474

INTRODUCTION

Wet air oxidation technology has been utilized for treatment of numerous industrial wastewaters. This paper presents the standard technological development path used to produce a continuous flow wet air oxidation system. A case study on the development and operation of two wet air oxidation systems is discussed. Experimental results from the bench scale batch testing and the continuous flow pilot plant wet air oxidation system are compared to the current operating full scale wet air oxidation systems.

The two wet air oxidation systems discussed are:

The wet air oxidation installation at Sterling Organics U.K., located in Dudley, Cramlington, Northumberland, England. The system is designed to treat para-aminophenol wastewater derived from the manufacturing of Paracetamol (i.e. acetaminophen).

The wet air oxidation installation at Global Environmental Services, located at the Knostrop Treatment Works, Leeds, England. The system is designed to handle various industrial wastewaters which are treated by a commercial wastewater treatment operation.

WET AIR OXIDATION PROCESS

Wet air oxidation refers to the aqueous phase oxidation of dissolved or suspended organic and inorganic substances at elevated temperatures and pressures [1].[1] The oxidation process requires oxygen from compressed air or pure oxygen as the oxidizing agent. The enhanced solubility of oxygen in aqueous solutions at elevated temperature and pressure provides a strong driving force for oxidation. Typical operating pressures for the wet air oxidation system vary from 100 to 3000 psig (690 to 20,700 KPa) and are dependent on the operating temperature. Commercial wet air oxidation systems generally operate at temperatures ranging from 150°C to 320 °C.

A basic flow scheme for a wet air oxidation system is provided in Figure 1. Oxidizable wastewater is pumped through the system using a positive displacement pump. Compressed air or oxygen is mixed with the pressurized wastewater. This mixture of gas and liquid is heated via heat exchange with the hot oxidized effluent. Typically, the wet air oxidation process can be designed for autothermal operation if the wastewaters contain greater than 15 g/l of chemical oxygen demand (COD). An external heater provides startup heat or additional heating when insufficient energy for autothermal operation is released by the oxidation reaction. The hot mixture flows into a bubble column reactor. A reactor provides the residence time for the oxidation reaction to occur. Retention time in the reactor varies from a few minutes to several hours depending on the type of wastewater and the treatment objective. Oxidized effluent flows from the reactor, through the process heat exchanger where it is cooled by exchanging heat with the incoming wastewater. Cooled oxidized effluent then passes through a pressure control valve into a separator where the oxidized liquid and non-condensable gases separate.

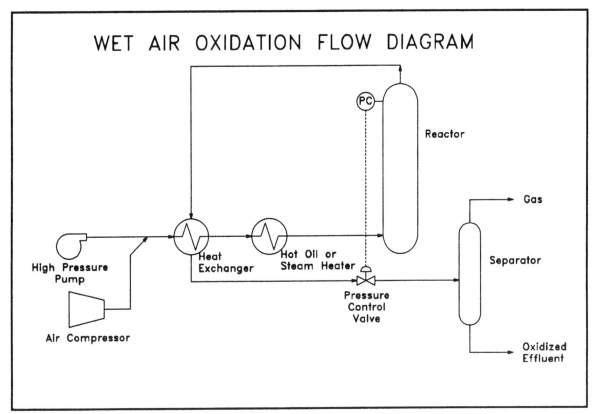

WET AIR OXIDATION FLOW DIAGRAM

Figure 1

WET AIR OXIDATION OF PARA-AMINOPHENOL

BATCH SCALE TESTING

Initial treatability testing of a wastewater from Sterling Organics U.K. began at the Zimpro facility in Rothschild, Wisconsin, in April of 1989. Bench scale shaking autoclave testing was performed on a wastewater containing phenol, para-aminophenol, and reduced sulfur. The wastewater was a by-product from the manufacturing of para-aminophenol, an intermediate used to make Paracetamol. At that time, the permitted disposal practice of this wastewater was direct discharge to the North Sea under license by the Ministry of Agriculture. In March of 1990, the Ministers of the North Sea States declared that industrial waste disposal into the North Sea would be phased out by the end of 1992. As a result, an investigation was initiated involving the use of the wet air oxidation technology to treat para-aminophenol wastewater.

Bench scale shaking autoclave tests are batch type tests used to determine the capability of the wet air oxidation technology for treatment of a particular wastewater. Figure 2 is a schematic of a batch type autoclave. Oxidation testing of the para-aminophenol wastewater was performed in titanium autoclaves having a capacity of 500 mls. The oxidations were performed at 3 different temperatures with a residence time of 60 minutes at temperature. Autoclaves were charged with 100 mls of waste, then pressurized with air to 800 psig and placed in a rocking mechanism which provides continuous agitation throughout the testing.

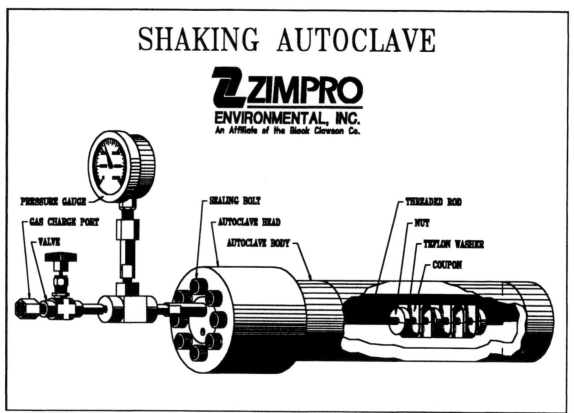

SHAKING AUTOCLAVE

ZIMPRO
ENVIRONMENTAL, INC.
An Affiliate of the Black Clawson Co.

PRESSURE GAUGE
GAS CHARGE PORT
VALVE
SEALING BOLT
AUTOCLAVE HEAD
AUTOCLAVE BODY
THREADED ROD
NUT
TEFLON WASHER
COUPON

Figure 2

Heat input was provided by electrical heating bands. After the oxidation test was conducted for the desired time, the autoclaves were rapidly cooled to room temperature in a water bath, then depressurized and the oxidized effluent was decanted and analyzed. Results of the oxidation testing are reported in Table I.

The test results show that the wet air oxidation technology would be expected to destroy over 98 percent of the phenol and greater than 99.9 percent of the para-aminophenol found in the wastewater. Destruction of COD ranged from 89.6 to 97.1 percent depending on the oxidation temperature. Complete oxidation of the reduced forms of sulfur did occur. The complete oxidation of reduced sulfur to sulfate is typical of the wet air oxidation process. The pH dropped during the oxidation process. The reduction in pH was attributed mostly to the formation of sulfuric acid (bisulfate) during the oxidation of the reduced sulfur present in the wastewater. Overall, the results from the bench scale batch testing indicated that the wet air oxidation process would be an excellent treatment technology for the para-aminophenol wastewater.

An extensive materials evaluation followed the bench scale oxidation study. The material of construction testing was performed in the batch autoclaves. Commercially available corrosion resistant alloys were tested. ASTM G-58 type corrosion coupons of various alloys were subjected to a wet air oxidation environment for over 1000 hours. Coupons were mounted in the autoclave utilizing a threaded support rod. Teflon washer placed between the coupons isolated them from contact with other metal and provided crevice areas for monitoring crevice corrosion.

TABLE I
BATCH SCALE SHAKING AUTOCLAVE RESULTS
FROM PARA-AMINOPHENOL WASTEWATER

Analysis	"As Received" Wastewater	200°C Oxidized Effluent	240°C Oxidized Effluent	280°C Oxidized Effluent
COD, g/l	52.1	5.4	3.9	1.5
COD Destruction, %	-	89.6	92.5	97.1
Phenol, mg/l	974	11.9	12.7	3.0
para-Aminophenol, mg/l	419	< 0.1	< 0.1	< 0.1
para-Aminophenol, Destruction, %	-	< 99.9	< 99.9	< 99.9
Sulfate-S, g/l	16	49.8	51.4	50.0
Sulfide-S, mg/l	11360	< 2.0	< 2.0	< 2.0
pH	10.1	2.1	2.1	1.7
Color, Alpha	22090	N.A	6739	N.A

Thread nuts were used to fasten the coupons in place and provide additional stress to the "as welded" cold bent coupons. Test coupons were placed into a autoclave containing the para-aminophenol wastewater and then pressurized with air. The testing temperature was 270 °C with the wastewater being changed every 48 hours.

Following the 1000 hour corrosion test, the various alloy coupons were evaluated for general and localized corrosion. The results indicated that 316-L stainless steel and alloy C-276 would corrode in the para-aminophenol wastewater during treatment. The 700 ppm of chloride found in the para-aminophenol wastewater caused transgranular stress corrosion cracking of 316-L. Alloy C-276 exhibited a high general rate of corrosion and showed evidence of crevice corrosion. Performance of the commercial pure titanium grade 2 and nickel containing titanium grade 12 was excellent. Neither of these alloy exhibited signs of general or localized forms of corrosion. The titanium material acceptably handled the chloride level and the large pH swing that occurred during the oxidation process. Based on the above testing titanium was selected as the material of construction for the wet air oxidation system treating para-aminophenol wastewater at 260 °C.

Additional bench scale oxidation testing was performed in July of 1989, when a portable bench scale stirred autoclave was shipped to Sterling Organics for on-site treatability studies. For a one month period of time, the stirred autoclave was used to demonstrate the effectiveness of the wet air oxidation process for treatment of the para-aminophenol wastewater. The results of the oxidation testing utilizing the stirred autoclave[2][2] were equivalent to those obtained from the shaking autoclave testing performed months earlier.

PILOT SCALE TESTING

In March of 1990 a continuous flow wet air oxidation test was conducted at Zimpro. Sixteen - 55 gallon drums of para-aminophenol wastewater were shipped to Zimpro from Sterling Organics U.K. The continuous flow wet air oxidation testing was performed in a pilot scale titanium system. The pilot plant system was capable of treating approximately 19 liters per hour with a reactor residence time of 60 minutes. The objectives of the testing were to demonstrate the acceptability of the wet air oxidation process in treating the para-aminophenol wastewater on a continuous flow basis, and to obtain design information.

The pilot plant wet air oxidation system design is a similar process flow scheme when compared to full scale wet air oxidation systems. A high pressure pump, air compressor, heat exchanger, reactor, effluent cooler, pressure control valves and separator are all present in the titanium pilot plant system.

Various oxidation test conditions were evaluated to determine the effects of residence time, temperature, and oxygen partial pressure. The wet air oxidation pilot plant testing was structured to provide operating data, which aids the process engineer in designing a full scale system. Valuable information such as: appropriate reactor residence time, reactor configuration, the amount of heat released during the oxidation, the heat transfer coefficients in the heat exchangers, scaling tendencies of the wastewater, off-gas composition, and effects of operating temperature, were all obtained by the continuous flow pilot scale testing. Also materials of construction choices were verified.

A partial list of analytical results obtained from the wet air oxidation pilot plant testing are displayed in Table II. The results were obtained from steady state operation of the pilot plant utilizing a nominal 60 minute reactor residence time. Results from both the 240°C and the 280°C pilot plant oxidations can be compared to the batch scale testing. The percent destruction of COD and para-aminophenol obtained by the two system compare well, demonstrating that the batch testing is a useful tool for evaluating wet air oxidation treatment capabilities. The destruction of para-aminophenol and reduced forms of sulfur obtained in the pilot plant testing were greater than 99 percent.

Data from the analyses of the off-gas from the pilot plant wet air oxidation system indicated that H_2S was not emitted. In addition, the SO_x and NO_x concentrations ranged from 13.7 ppm to 3.5 ppm and 0.000 ppm to 0.012 ppm, respectively.

The major conclusions obtained from the pilot plant wet air oxidation study were:

1) Wet air oxidation could effectively treat the para-aminophenol wastewater at 260°C with a one hour reactor residence time.

2) Scaling tendencies of the wastewater were minimal and precipitation of salts due to solubility limitations could be restrained with appropriate temperature and pressure control.

3) Titanium was the appropriate material of construction.

TABLE II
CONTINUOUS FLOW PILOT PLANT ANALYTICAL RESULTS
FROM PARA-AMINOPHENOL WASTEWATER

Analysis	"As Received" Wastewater	240°C Oxidized Effluent	260°C Oxidized Effluent	280°C Oxidized Effluent
COD, g/l	71.9	4.43	2.68	2.46
COD Destruction, %	-	93.8	96.3	96.5
Total Phenols, mg/l	846	101	61.9	16.9
para-Aminophenol, mg/l	5130	0.38	0.07	0.04
para-Aminophenol, Destruction, %	-	< 99.99	< 99.99	< 99.99
Sulfate-S, g/l	12.2	55.4	60.6	58.5
Sulfide-S, g/l	5.9	< 0.064	< 0.064	< 0.064
Thiosulfate, g/l	71.3	< 0.45	< 0.45	< 0.45
pH	10.0	1.9	1.8	1.65
General Appearance	Very Dark Brown	Very Dark Brown	Light Yellow Color	Very Light Yellow

FULL SCALE SYSTEM

Based on the acceptable treatment results obtained from the pilot plant testing, a full scale wet air oxidation system for treatment of the para-aminophenol wastewater was procured by Sterling Organics from Zimpro. Commissioning and a final acceptance testing took place between August 15, 1992 and January 15, 1993. The wet air oxidation unit was designed to process 112 liters/minute (30 gpm) of para-aminophenol wastewater with an oxygen uptake of 80 g/l. The reactor operates at 260°C (500°F) and 10.3 MPa (1500 psig) with a one hour hydraulic residence time. All high temperature/high pressure materials in contact with the process liquid were manufactured from titanium grade 2 or 12. The flow diagram for the full scale system is essentially the same as shown in Figure 1.

Several minor processing problems occurred during the initial startup of the wet air oxidation system. After several days of continuous operation a grey polymeric solid formed in the heat exchanger inlet piping. Formation of these solids eventually restricted flow in the line. Analysis of the solids indicated a organic material containing 50 percent reduced sulfur. The formation of these solids was enhanced in the presence of air. On further inspection it was determined that slightly higher temperatures would inhibit the formation of the polymeric material. The mixing point of waste and air was modified so that the combination of the two streams would occur in the hot area of the heat exchanger. After the modification, the formation of the gray solid in the inlet line has not occurred.

At one point during the commissioning of the system, an inspection of the reactor revealed elemental sulfur deposits. The cause of the deposits was traced to an upset condition where air flow was lost due to an air compressor shutdown. For a 10 to 20 minute time period, the wet air oxidation system continued to process wastewater in the absence of oxygen. The oxygen deficient conditions caused the formation of the elemental sulfur. Following this incident, Sterling Organics modified the computer control system so that during upset conditions the wet air oxidation system would be placed on water. Since this modification, sulfur deposition has not been a problem.

Analytical results from the acceptance test are listed in Table III. These results indicate that the full scale wet air oxidation system destroyed over 99.99 percent of both the para-aminophenol and reduced sulfur. The oxidation process also destroyed over 96 percent of the COD and 97 percent of the phenol in the wastewater. Both the levels of destruction and effluent quality closely match the results obtained during pilot plant and bench scale testing.

Performance of the Sterling Organics wet air oxidation system exceeds the regulated effluent requirements. The full scale wet air oxidation system at Sterling Organics was placed into on-line operation following the completion of the acceptance test on January, 1993. As of January, 1994, the system has process over 35,000 cubic meters (9 million gallons) of para-aminophenol wastewater. The para-aminophenol wastewater is typically produced at a rate of 56 to 70 liters per minute. Since the acceptance of the wet air oxidation system on January, 1993, all para-aminophenol wastewater generated has been treated by the wet air oxidation system. Total run hours on the process air compressor indicate that the wet air oxidation system has had a 90.2 percent availability for the first year of operation. The wet air oxidation system is designed to operate with an 80 percent availability.

TABLE III

FULL SCALE ACCEPTANCE TEST ANALYTICAL RESULTS
FROM PARA-AMINOPHENOl WASTEWATER

Analysis	para-aminophenol Wastewater	Design Treatment Quality	260°C Oxidized Effluent
COD, g/l	69.1	< 4.0	2.6
COD Destruction, %	-	-	96.1
Phenol, mg/l	774	<100	18
para-Aminophenol, mg/l	6950	< 0.5	0.067
para-Aminophenol, Destruction, %	-	-	< 99.99
Sulfate-S, g/l	10.2	56.3	49.1

During long term continuous operation it was found that heat transfer capabilities of the heat exchanger decrease over time. As part of the standard operating procedure Sterling Organics switches from para-aminophenol wastewater feed to recycled oxidized effluent, which removes scale build up in the tubes of the heat exchanger. Cleaning of the heat exchanger is performed easily and economically without cooling the system down. The feed switch over is done once a week and takes approximately 2 hours.

WET AIR OXIDATION OF VARIOUS WASTES FOR COMMERCIAL TREATMENT

OBJECTIVES AND DESCRIPTION OF THE WET AIR OXIDATION SYSTEM

Global Environmental Services operates a 285 liters/minute merchant waste treatment facility located on the Knostrop Sewage works site in Leeds, England. The treatment facility was designed to treat a wide range of industrial wastewaters. In order to expand the overall treatment capabilities of the facility, Global Environmental procured a laboratory bench scale autoclave testing system and a full scale wet air oxidation system from Zimpro. The intent of Global Environmental Services was to use the bench scale equipment to pre-screen wastewaters produce by outside industrial facilities. Then, after testing, Global Environmental Services offers wastewater treatment services for those wastewaters that are amenable to wet air oxidation treatment.

Bench scale oxidation tests on over 20 different wastewaters were performed by Global Environmental Services at their Knostrop Treatment Works. The bench scale testing began in October of 1992, using the equipment procured from Zimpro. Analytical results from the autoclave bench scale testing were used to aid in selecting 6 (six) different wastewaters to be treated in an acceptance test for the full scale wet air oxidation system. Acceptance of the wet air oxidation system was based on results obtained from the batch scale testing and the full scale system. Acceptable performance of the full scale wet air oxidation system meant obtaining treatment comparable to that obtained by the bench scale autoclave testing, operating at equivalent test parameters (eg. COD destruction, BOD_5/COD ratio, respiration inhibition reduction). Another major treatment requirement for acceptance of the full scale system was to demonstrate the ability of the system to operate at the design flow rate, temperature, and pressure.

The full scale wet air oxidation system is capable of treating wastewaters containing up to 32 g/l of COD (oxygen uptake) at a flow rate of 38 liters/minute (10 gpm) with a nominal hydraulic residence time in the reactor of one hour. The maximum operating temperature is 280°C (536°F). The wet air oxidation system was designed with adjustable feed and air flow rates to allow for a large operating window. Dilution water addition was also a process option to allow for treatment of higher strength wastewaters. Grade 2 and 12 titanium were the materials of construction for the high temperature, high pressure components. Titanium offers the best corrosion resistance for a system treating a large variety of wastewaters.

WET AIR OXIDATION TESTING

Side by side full scale and batch scale wet air oxidation tests were performed during the acceptance testing of the full scale system. The test parameters for each system were set to produce equivalent wet air oxidation environments so a direct comparison of the treatment results was possible. The start of the acceptance test began in July of 1993.

Comparative results from oxidation testing from both the bench scale and full scale system are listed in Tables IV, V, and VI. Each system utilized the same test parameters (oxygen partial pressure, temperature, residence time, wastewater feed). High levels of COD destruction were obtained on the six wastewaters tested. The reduction in COD obtained from the full scale system ranged from 67 to 77 percent, depending on the wastewater. COD destruction percentages for the two systems treating the six different wastewaters were comparable. Four of the COD destruction percentages from the full scale system were at least 90 percent of the results obtained in the bench scale autoclave testing. The overall average of COD destruction efficiency of the two systems shows that the full scale system achieved 90 percent of that obtained in the shaking autoclave. The slight variations in the treatment capabilities of the two systems is generally explained by the longer heat-up time in the bench scale system.

TABLE IV
WET AIR OXIDATION TESTING RESULT
COD DATA FROM FULL SCALE AND BENCH SCALE TESTING

Waste Type	COD (mg/l)			COD Reduction		
	Feed	Full Scale Unit Effluent	Bench Scale Effluent	Full Scale Unit Reduction (%)	Bench Scale Reduction (%)	Ratio Full Scale/ Autoclave
Phenolic Wastewater	35400	10600	7980	70.1	77.5	90
Organic Nitrogen Wastewater	24400	7930	5990	67.5	75.4	90
Organic Nitrogen Wastewater	7750	2510	1450	67.6	81.3	83
Caustic Scrubber	36900	12000	10600	67.5	71.3	95
Chlorinated Organic Amine Wastewater	35500	11200	6390	68.5	82.0	84
Pesticide Wastewater	22500	5170	4500	77.0	80.0	96
Average						90

The BOD$_5$/COD ratios obtained from the full scale system, treating the six different wastes, ranged from 0.27 to 0.89, indicating that the remaining COD would be amendable to further biological treatment. Comparison of the BOD$_5$/COD ratio between the two systems indicate that the bench scale and full scale performance was within 10 percent of each other. The overall average of the six wastewaters tested showed the BOD$_5$/COD ratio obtained from the two systems to be within 2 percent of each other.

The results indicate a large reduction in toxicity was obtained by the wet air oxidation system. The respiration inhibition (RI) reduction results obtained by the full scale wet air oxidation system treating the six different wastewaters ranged from 38 to 98.6 percent. The comparison of RI reduction between the full scale and bench scale systems showed that five out of the six samples had full scale/autoclave ratio of 90 percent or greater. The overall average of the six wastewaters tested showed the RI reduction obtained from the two systems to be within 3 percent of each other.

TABLE V
WET AIR OXIDATION TESTING RESULT
BOD$_5$ DATA FROM FULL SCALE AND BENCH SCALE TESTING

Waste Type	BOD$_5$ (mg/l)		BOD$_5$/COD Ratios		
	Full Scale Unit Effluent	Bench Scale Effluent	Full Scale Unit Ratio	Bench Scale Ratio	Ratio (Full scale/ Autoclave)
Phenolic Wastewater	8400	6550	0.79	0.82	96
Organic Nitrogen Wastewater	2190	1820	0.27	0.30	91
Organic Nitrogen Wastewater	1050	641	0.41	0.44	95
Caustic Scrubber	8000	7390	0.66	0.69	96
Chlorinated Organic Amine Wastewater	7670	4150	0.68	0.64	105
Pesticide Wastewater	4650	3210	0.89	0.71	126
Average					102

TABLE VI
WET AIR OXIDATION TESTING RESULT
RI DATA FROM FULL SCALE AND BENCH SCALE TESTING

Waste Type	RI Data			RI Reduction		
	Feed	Full Scale Unit Effluent	Bench Scale Effluent	Full Scale Unit Reduction (%)	Bench Scale Reduction (%)	Ratio Full Scale/ Autoclave
Phenolic Wastewater	50	4	4	92.0	92.0	100
Organic Nitrogen Wastewater	14	4	5	71.4	64.2	111
Organic Nitrogen Wastewater	15	6.3	6.1	58.0	59.3	98
Caustic Scrubber	36	6	7.4	83.3	79.4	105
Chlorinated Organic Amine Wastewater	3250	45	50	98.6	98.5	100
Pesticide Wastewater	50	31	28	38.0	44.0	86
Average						100

CONTINUOUS OPERATION

Following the suitable performance of the full scale system during the acceptance test, Global Environmental presented the treatment data to the Regulatory Authorities with intention of obtaining permits to operate the wet air oxidation system for commercial purposes.

The approval of the operating permits was completed on November 8, 1993. The wet air oxidation unit is now operating continuously and acceptably. Initial deposits of scale in the process heat exchanger were observed during startup of the system on hard plant water, but scaling has been avoided since, by using oxidized effluent for startup and dilution water.

CONCLUSIONS

The wet air oxidation process has been acceptably applied to the treatment of a para-aminophenol wastewater and as a pretreatment system for a commercial wastewater treater.

The high level of performance demonstrated by the Sterling Organics wet air oxidation system indicates that the wet air oxidation technology is a reliable means of wastewater treatment. This wet air oxidation system has operated with over a 90 percent availability.

Similar treatment efficiency was obtained by the bench scale, pilot scale and full scales systems. The comparable treatment results consequently lead to accurate design scale up of full scale systems using bench scale and pilot scale data.

REFERENCES

1. Schaefer, P.T., "Consider Wet Oxidation, Hydrocarbon Processing", October 1981

2. David N. Preece and Dr.William M. Copa et. al. March 1993. "Wet Air Oxidation of Waste Water," The Industrial Waste Water Treatment Conference, United Kingdom, Sterling Organics and Zimpro Environmental.

Milton Keynes UK
Ingram Content Group UK Ltd.
UKHW050608161024
449569UK00046B/1529

9 781566 764896